WATER MANAGEMENT *in* AFRICA *and the* MIDDLE EAST

CHALLENGES *and* OPPORTUNITIES

EDITED BY
EGLAL RACHED, EVA RATHGEBER, AND DAVID BROOKS

INTERNATIONAL DEVELOPMENT RESEARCH CENTRE
Ottawa • Cairo • Dakar • Johannesburg • Montevideo
Nairobi • New Delhi • Singapore

Published by the International Development Research Centre
PO Box 8500, Ottawa, ON, Canada K1G 3H9

May 1996

Legal Deposit: 2nd quarter 1996
National Library of Canada

A microfiche edition is available

To order, please contact IDRC Books.

Mail: PO Box 8500, Ottawa, ON, Canada K1G 3H9
Phone: (613) 236-6163 ext. 2087
E-mail: order@idrc.ca

Our catalogue may be consulted on Internet.

Gopher: gopher.idrc.ca
World-Wide Web: http://www.idrc.ca

ISBN 0-88936-804-X

Contents

Part III. Special issues

PREFACE

This book is the product of an International Development Research Centre (IDRC) workshop to take stock of the present situation of water management in Africa and the Middle East. The workshop was held in December 1994 in Cairo, Egypt. Twelve formal presentations were given: three overview papers on key water issues; seven subregional papers on key water issues; and two transsectoral papers, one on nongovernmental organizations' perspectives and the other on sectoral water allocation.

Each presentation was followed by lively discussions, and a day was devoted to the identification of priorities for research and development. Two additional papers have been included in this publication, the first dealing with women's roles as water users and managers; the second with, grassroots utilization of water resources. The full report of the conclusions and recommendations from the workshop has not been included in this book, but is available from IDRC's Middle East Regional Office in Cairo upon request.

Although some papers were edited only for style, others had to be substantially revised. Limitations of time and communication did not permit all the authors to respond to a number of specific questions formulated by the editors. In a number of cases, references were vague or incomplete and had to be removed. However, every effort was made to neither change the meaning of the text nor modify the author's approach.

Eglal Rached
Eva Rathgeber
David Brooks
IDRC

ACKNOWLEDGMENTS

We would like to express our special thanks to Sahar Kamel and Maryse Rabbat for their unsparing efforts to make this workshop a success and for their sustained secretarial assistance and to Brenda Lee Wilson, who contributed a lot of time and thoughtful comments to the editing. Many thanks also go to other IDRC staff — Rosa Ongeso and Muthoni Mwangi, Nairobi, and Ray Vander Zaag and David Ofoumon, Ottawa, for editorial assistance; Flora Shiroya, Nairobi, for bibliographic assistance; and Imelda Wasike, Nairobi, for secretarial assistance. Finally, we thank the text editors, Wilma Fraser and Josephine Mwasi.

Introduction

Eglal Rached
Senior Program Specialist, Middle East and North Africa Regional Office, IDRC
Cairo, Egypt

It is increasingly apparent that a huge international water problem will be looming over development in the coming decades. Vivid attention to this situation was recently drawn by Dr. Ismail Serageldin, Vice President of the World Bank, who warned that "unless current trends are reversed, we will have a major water crisis around the planet."

Nowhere more than in Africa and the Middle East (A&ME) is water likely to become the most critical resource issue and the most limiting input to food security and economic and social development. The countries with the most limited water resources are in the Middle East and northern and southern Africa. Despite the common impression of Africa as a jungle, 54% of the continent is arid to semi-arid, and only 14% is humid to very humid. The remaining 31% has good rainfall. According to estimates, Kenya's water supply will shrink by 50% and Nigeria's by 40% in under a decade. The situation in the Middle East and North Africa is even worse: only 19% of the land surface receives good rainfall; with the partial exception of Turkey, the whole region is arid to semi-arid.

In A&ME, water supply for agricultural, domestic, and industrial use, as well as for environmental use (rivers, habitat preservation, transportation, fishing), has kept pace neither with population growth, which is the fastest in the world, nor with economic growth, which is booming in many countries in the region. As a result, by 2025, the amount of water available per person in the Middle East and North Africa is expected to drop by 80%, in a single lifetime, from 3430 m^3 to 667 m^3. New or additional sources of water are becoming scarce and more expensive to develop, and waste products are increasingly contaminating available surface and groundwater sources. Because of its scarcity, water is also becoming one of the most important causes of internal and international conflicts.

Although the crisis threatens all of A&ME, individual countries and specific subregions face different situations, constraints, and opportunities. These are related to the wide range of geography and topography, climatic and hydrological conditions, and institutional, political, economic, and cultural situations in the region. Few of these variables are well understood, and all of them need to be carefully assessed before viable and sustainable solutions, tailored to each situation, can be identified.

Conscious that any move toward resolution must be based on what people on the ground know, can do, and want to do, the International Development Research Centre (IDRC) convened, in December 1994, a Pan-African workshop on water management for researchers, practitioners, nongovernmental organizations (NGOs), and policymakers. The purpose of the meeting was to take stock of the present situation and to get a better understanding of the opinions and priorities that should be taken into consideration when setting strategies for sustainable water management. It was anticipated that this assessment would provide the basis for designing a priority research and development agenda that would strengthen efforts of those already involved in research, advocacy, development, and environmental water activities.

The IDRC initiative aimed to identify research priorities and options from the perspective of users' and communities' needs and priorities. Three areas of emphasis were singled out:

- power to access and use water, including sources of stress involving water supply, abuse of water resources, resulting conflicts, and processes to resolve such conflicts;
- demand management and water conservation, including sectoral water allocation and efficient demand management to improve availability and quality; and
- future options for improving water supply, including alternative technologies and analyses of the potential for efficient and equitable management of water supply by communities and local groups.

The workshop had 12 formal presentations: three overview papers on key water issues; seven subregional papers; and two additional papers, one reviewing NGO perspectives on water and the other on choices and impacts of sectoral water allocation in Africa. Two papers were contributed later to highlight specific issues; one paper was on women's roles and constraints as water users and managers, and the other was on grass-roots utilization of water resources. Each presentation was followed by an open discussion. Three task groups then identified priorities for research and development. IDRC is publishing the proceedings of the meeting to widely disseminate the accumulated knowledge, to facilitate further sharing of opinions and debate among the various actors involved in water management, and to assist in formulation of priority research and development areas for A&ME.

Regional characteristics

Northern Africa and the Middle East are the most water-scarce regions in A&ME. Countries have consequently invested heavily in the mobilization of their water resources. Between 1970 and 1990, large-scale systems for water supply and conveyance systems were built, and the total irrigated area increased dramatically. The improved water infrastructure stimulated economic growth and reduced somewhat the damaging effects of droughts. However, many of these countries are entering a critical phase. Almost all the states of the Arabian Peninsula, plus Jordan, Israel, and Libya, already consume much more water than their annual renewable supply. Egypt, Sudan, Morocco, Tunisia, and Syria are fast approaching this situation. At the same time, opportunities for building dams and abstracting groundwater are becoming more scarce and more expensive, and widespread pollution and salinity of surface and groundwater sources are further reducing available supplies. Under these conditions, the most important course of action shifts to the management of demand, particularly through the economic valuation of the resource and more efficient allocation of water among different sectors. Given that most of these countries devote 60–90% of their water to irrigation, water conservation in the agricultural sector is considered one of the most important future options. The conjunctive use of rain and groundwater resources to improve rainfed agriculture is another promising option.

On the supply side, cost-effective options have been almost completely exhausted. In addition, most, if not all, have serious social and environmental problems that in the past went unsolved. Water recycling is commonly practiced in the region for irrigation but, if not very carefully controlled, entails both human health and soil-quality problems. A number of water sources are being explored, including some high-cost, highly centralized options such as sea and brackish water desalinization, water imports, and cloud seeding.

Fortunately, low-cost, decentralized options are also beginning to attract the attention of water managers. Of particular interest is the improvement of rainwater or flood-water harvesting, either for supplemental irrigation or for aquifer recharge. Rainwater collected from rooftops can also be an important source of potable water. Agricultural techniques appropriate for water scarcity, such as use of poor-quality or saline water for irrigation, are also gaining attention, though here, too, careful technique selection is required to protect aquifers and preserve soil quality. In addition to cost effectiveness, water-harvesting techniques have several advantages, including simplicity and suitability for community-level planning, funding, and overall management (and therefore for sustainability); ability to provide water to isolated communities; erosion control; soil-harvesting potential; and flood control.

Rural and pastoral communities in North Africa and the Middle East have for centuries practiced water-harvesting techniques, conjunctive use of surface water and groundwater for dryland irrigation, and rainfed agriculture. Their knowledge of, and experience with, these and other interesting examples of local water resource management have considerable potential for transfer within North Africa and the Middle East or to other parts of Africa.

In contrast to North Africa and the Middle East, rivers and streams are the main sources of water for most of East and West Africa; underground and flood-water resources are largely unexplored. Only about 45% of the irrigation potential is utilized in the Sahel region; there is, therefore, considerable scope for irrigation extension. However, a number of constraints hamper the development of these resources. Some of these restraints are technical, but most are institutional, socioeconomic, and cultural. Significantly, national governments have emphasized large-scale schemes and the installation of water-supply systems, rather than their maintenance. Today, these governments find themselves strapped by limited finances and technical staff, administrative backlogs, inappropriate or expensive institutional structures, and poor legislative support. Because of high development costs and the plethora of management problems resulting from big irrigation schemes, large-scale supply options have lost favour. An approach that transfers management of water to local communities and builds on local knowledge, institutions, and solutions for better water management has considerable potential. The greatest challenges are to design and implement low-cost technologies for improving water supply and management at the local level.

Some of East Africa and much of West Africa fall within the humid tropics, where abundant water resources are available. There too, despite this availability, numerous water-related problems exist. Despite the proclamation by the United Nations in 1980 of the International Drinking Water Supply and Sanitation Decade (1981–90), aimed at supplying 500 000 people with safe drinking water and adequate sanitation by 1990, more than half of the African population has neither safe drinking water nor sanitation. As in the drier areas of sub-Saharan Africa, irrigation is largely neglected in the subregion and a significant potential for expansion remains. With a rapidly expanding food gap, particularly in cereals, the expansion of the irrigated sector is becoming urgent. Because of economic and social costs associated with large-scale irrigation systems, the potential for small-scale, community-based irrigation systems is considered the greatest, especially if combined with water harvesting and improved water efficiency.

The situation in southern Africa is, in many respects, similar to that of North Africa and the Middle East. Water is in limited supply in most countries of the subregion. Most watersheds are already tapped by large-scale abstraction schemes, and plans are under way to tap the remaining ones. Botswana and Namibia have already reached their internal water-supply limits. The same applies to Zimbabwe, where, in September 1995, a water-shortage alarm was issued when the lack of water throughout the country had become critical. Major dams, including those supplying large towns and cities, were only 18% full, and 11% of the boreholes and 20% of the deep wells in rural areas had dried up. Despite a major operation to drill boreholes all over the country, the rural areas, where most of Zimbabwe's 10 million people live, were badly affected. Crop production was slashed by the failure of the rains in the summer growing season, and energy was in critically short supply as a result of the lowest rainfall in the river catchment since 1907. Government officials warned that "if consumption continues at the current level, we are likely to run out of water before the next rainy period."

Water stress is particularly important in South Africa because of the size of its population and economy. Key policy options are similar to those of the Middle East and North Africa and include demand-side management, the decentralization of water management to communities and to private local structures, and water harvesting for small-scale agriculture.

Some overriding issues for debate

Early in the workshop it was pointed out that a sharp difference exists between the immediate water problems in North Africa and the Middle East, on the one hand, and those in sub-Saharan Africa (with the exception of South Africa) on the other. In North Africa, the Middle East, and South Africa, the urgent need is to divert water from agricultural use to domestic, municipal, industrial, and environmental uses for two reasons: the demand for freshwater has exceeded the supply for many years, and agriculture is providing a declining share of the gross national product and employment. The reverse is true of humid, subhumid, and even Sahelian Africa, where irrigation has remained largely unexploited and where more water could be productively and equitably applied to agriculture. Expansion of irrigation would also reduce vulnerability to drought and increase food security.

Different views were expressed on the issue of large- and small-scale irrigation and dam projects. Although the majority of participants felt that small-scale, low-cost, locally applied technologies offered considerably greater potential than had been recognized, and with fewer disadvantages, proponents of large irrigation projects believed that there is no way for such projects to be avoided if rapid and substantial agricultural production increases are to be achieved. The challenge is to do them right when there are no feasible small-scale alternatives. There was, however, consensus on the need to satisfy human needs for food, water, and economic opportunities while reducing environmental degradation.

The analogy between the energy and water sectors was presented and discussed at some length. Both resources are generally oversupplied relative to demand, underpriced relative to their intrinsic and economic values, and governed by institutions geared to increase supply rather than manage demand. The analogy may be particularly useful for Africa, where a relatively high proportion of energy comes from hydroelectricity and many major water projects are designed for both electricity generation and irrigation.

The issue of water pricing raised considerable debate. Underpricing of water, which is practiced today in most countries of the region, allows low-value users such as agriculture (which accounts for 88% of end-uses overall) to consume large quantities of water and to use it wastefully. The result is depletion, degradation, and physical and economic losses. Underpricing also results in unreliable service, unwillingness to pay, and decline in capacity to provide services. Higher income people and industry are generally able to cope with this situation, but poor people and local industry cannot. The result is that the vast majority of poor people are denied adequate water and sanitation services.

On the other hand, the introduction of water pricing is not an easy task. Complex social, cultural, political, and economic factors will influence water availability, allocation, and use. As well, pricing may have inequitable impacts on different classes and sectors. Although some participants suggested that other avenues (for instance, improvement of water-management strategies at the farm level, or improvements in operation and maintenance of irrigation schemes) should be considered for water conservation, others suggested that water pricing was unavoidable and not as unpopular as one might think, provided reliable services and quality are assured. Indeed, in many places where people buy water from vendors, they pay a high price for it. No one at the workshop advocated a purely market solution for water. The participants agreed that some constraints were necessary to ensure that the market operated efficiently and equitably.

Institutions as a priority

Institutional issues emerged throughout the workshop as being at the root of water-management problems. The collection, treatment, distribution, and disposal of water are generally overcentralized, with little community participation. All too often, planners don't consult the people most directly involved in deciding how water should be distributed. Planners also fail to ensure that users, who are customarily expected to share in the maintenance of facilities, bear such responsibilities in proportion to their benefits from the system. As a result, government agencies are overextended, lack the proper incentive structure, and are unable to provide high-quality services. In addition, water-management responsibilities are fragmented, with little regard for conflicts or complementarities among social, economic, and environmental objectives. Finally, because of underpricing, water utilities may be unable to recover their own costs, which leads to a vicious cycle of inadequate maintenance, poor services, unwillingness to pay even nominal charges, and then still lower revenue, and so on.

Participants suggested that key changes require the separation of water management from delivery. Policy, planning, and regulatory functions should be separated from engineering, construction, and operational activities at each level of government. Operations should be assigned to specialist agencies where appropriate, and responsibility should be decentralized to the extent possible. Without assuming that local management is inherently equitable and efficient, participants nevertheless felt that the role of government should be primarily to create an enabling environment in terms of policy, legislation, institutions, and public awareness and to generate local initiatives and individual incentives. Moreover, these are just the functions least likely to be effectively delivered at local levels.

Participants from NGOs and advocacy organizations put particular emphasis on the need to involve communities in the development, implementation, and management of water projects. A number of presentations highlighted the fact that, too often, community participation is used by authorities to download the financial and labour costs of construction and maintenance of waterworks. It was pointed out that *community participation* was in itself misleading because it often did not take into account existing power structures within the community and frequently ignored the specific interests, roles, responsibilities, and needs of women in water management and utilization.

Over the next decades, the single most important consideration in water management will have to be that of institutional design — developing a set of rules that users, suppliers, and policymakers understand, agree upon, and are willing to follow. Organizational blueprints do not exist, nor are they likely to be readily adopted. Instead, within broad frameworks set by government, users and suppliers need to design their own institutions, matched to their particular set of physical, economic, and cultural conditions. At the same time, it is also important to keep in mind that decentralization and market systems have their own problems: perverse incentives also face private firms and local management institutions responsible for systems operation and maintenance.

Another issue of growing important is the international conflicts over water quantity and (recently) quality. Rights to use or harvest are creating difficult issues for governments throughout the region. Participants stressed the need to quantify transborder impacts of water use and disposal. High priority must be accorded to improved methods for conflict resolution and joint management, whether these involve two communities on opposite sides of a border or entire nations depending upon the same watercourse or aquifer.

The above list is by no means exhaustive. Participants discussed numerous technical, socioeconomic, and political issues facing researchers in A&ME, which included the following:

- How can better integration be achieved between land and water management, ensuring sustainability of flora, fauna, and habitat?
- How do existing tenure and property arrangements, including common property concerns, affect the way water is managed and services delivered?
- What is the appropriate role for and the most effective ways of incorporating indigenous knowledge into water decision-making and management?

Although consensus was not reached on solutions and the types of interventions required, questions and differences were clearly articulated and point to the need for more rigorous examination of the substance and specifics behind the issues. In some cases, enough information is already available for improved water policies and practices. In many others, however, additional research is required before agreements can be reached on appropriate priorities and strategies.

Finally, it is important to stress that the majority of the presentations in this publication have been prepared by scientists living and working in A&ME who deal on a daily basis with the looming water crisis. They are therefore the ones who are in the best position to suggest alternative solutions or processes that might lead to possible solutions.

For its part, IDRC will continue to place a high priority on the support of applied research on water management in A&ME. As in the past, the great bulk of that research will be proposed, implemented, and carried forward into policy proposals by researchers from the region.

PART I

Concepts

Demand-side Management, Conservation, and Efficiency in the Use of Africa's Water Resources

Geoffrey Stiles

Energy Conservation Advisor, SADC Energy Management Project, Harare, Zimbabwe

Introduction

This paper provides a conceptual framework for improved demand management of water resources in Africa based on the experience of demand-side management (DSM) of energy resources in Canada and the United States. (The difference between demand management and DSM is explained in the section "Demand-side management for water: a basic paradigm.") For examples of efficient water use and for an overall view of water management in Africa, the paper focuses on sub-Saharan Africa. North Africa is discussed only in relation to the experience of irrigation on the Nile. The paper also concentrates on the use of water for human consumption and excludes discussion of water's role as an ecological entity (for example, its use for habitat preservation).

Experience from energy-demand management provides a useful analogy to demand management of water resources, even if incomplete. As Brooks (1994) pointed out, the analogy works despite its imperfections: water is often oversupplied relative to demand, generally underpriced relative to its intrinsic and economic values, and governed by "institutions geared to augment supply rather than to manage demand."

Indeed, the analogy may be particularly useful for Africa, where a relatively high proportion of energy comes from water sources (that is, hydroelectricity) and many major water projects are dual-purpose (they provide water both for consumptive uses, such as irrigation, and for nonconsumptive uses, such as electricity generation).

There are a number of parallels in the use of water and energy resources that should be noted:

- Water, as with energy, is commonly transmitted from its point of collection (generation) to its point of use, and it is in the transmission process that substantial losses (inefficiencies) are typically incurred.

Moreover, for both resources, means of transmission are extremely variable. In the case of water, they range from fairly simple manual technologies for gravity-based irrigation canals to extremely complex capital-intensive transmission by pumps and pipelines. In water, as in energy transmission, the transmitter and the user are faced with important technological choices that may affect overall system efficiency.

- The end-uses of water, as with those of energy, typically require some form of transformation technology. This can range from a simple faucet or showerhead in a domestic reticulation system, to canals and spray irrigators in agriculture, to boilers and purifiers in industry. Although the parallel to energy is incomplete because the transformation is not necessarily thermodynamic (Brooks 1994), it is close: water must often be changed in form, as to steam or hot water, or in force or pressure before it can be used effectively.
- Water, as with energy, is a critical resource and lends itself to the imposition of control and authority. This point was made by Wittfogel (1957) in his classic study of "hydraulic civilization." Wittfogel was preoccupied primarily with control over the supply of water, but it is equally important to consider how demand might be controlled and what role water providers can take in efforts to improve demand management of water.

My principal source of information on the energy side is the DSM experience of power utilities in North America and Europe. Energy DSM provides special insights into how user needs can influence the resource management process and resource management authorities in particular, a process that can lead to a profound change in financial management and investment patterns. There are also cases from North America where DSM has been specifically applied to the management of water.

Water use and demand management in Africa

Water use in Africa, as in other developing regions, is dominated by agriculture. Indeed, Africa has the highest percentage use by agriculture of any major region and, conversely, the lowest percentage of domestic or industrial uses (Xie et al. 1993). As a rule, countries increase their relative share of domestic and industrial use as they become more urbanized and more "modernized" or as income levels increase (World Bank 1992). Thus, the dominance of agricultural end-uses in

Africa indicates that domestic and industrial supply systems are relatively undeveloped, with a few notable exceptions like South Africa, Egypt, and Zimbabwe. The extremely rapid increase in Africa's urban population has done little to change this, and most rural–urban migrants end up in informal settlements with dangerously inadequate water supplies. There is also little capital available to pay for the further development of industrial–commercial water supplies in burgeoning African cities. Because of this, Africa has been the target of a number of massive and relatively well documented efforts at water management, particularly those involving riverbasin transfers. Africa also provides examples of the use of water-efficient technologies and management systems in all sectors.

Water end-uses

Commercial or market-based agricultural use

It is estimated that 88% of water in Africa is used for agriculture (Xie et al. 1993). The area under irrigation is about 5.3×10^6 ha, of which 2.6×10^6 ha (53%) is for commercial or large-scale agriculture (Wyss 1991). Commercial irrigation schemes include irrigation from pumped groundwater, irrigation from large dams, and various kinds of river transfers to irrigation systems. The balance, mostly traditional or small-scale irrigation, consists of private flood, swamp, surface, and low-lift irrigation (Wyss 1991). Most river-based schemes are used for capital-intensive farming, with production primarily for the market, although a surprisingly large amount is used for small-scale farms in countries such as Sudan (OECD 1984; World Bank 1990).

Although these figures give the impression that irrigation is relatively common in African agriculture, only 5% of total crop land is, in fact, irrigated (Grainger 1990; MacLean and Voss, this volume), compared with 29% for Asia. This is to some extent a function of the peculiar geography of Africa, which includes inaccessible aquifers in West Africa, making groundwater irrigation extremely expensive, and unreliable surface-water flows elsewhere, making other kinds of irrigation highly risky (Grainger 1990). Rivers in many parts of the continent run dry for long periods of the year because they originate in drought-affected highlands. The impact of these erratic flows is increased by devegetation and the resultant low capacity of the soils to absorb water. It is in these relatively marginal and seasonally semi-arid conditions that a majority of the continent's farmers, both commercial and traditional, attempt to make a livelihood.

Despite these adverse conditions, there is great potential for increasing irrigation in sub-Saharan Africa (MacLean and Voss, this volume). The irrigation potential of this subregion is estimated at $20–33 \times 10^6$ ha (Elahl and Khushalani 1991; Bryant 1994), although much of this is land that could only be irrigated after substantial capital investments. The additional irrigation that might be achieved, given market constraints and limited capital, is probably less than half this amount.

By far the largest amount of large-scale commercial irrigation for farming is in Sudan, where a number of projects have had a massive impact on agricultural production. Good examples are the Gezira and Rahad schemes, which, together, cover several thousand square kilometres. The overall environmental and economic impacts of such schemes are extremely contentious, and, despite major increases in production of commercial crops, the irrigation systems feeding the main growing areas have been subject to erratic management and poor maintenance. As a result, crop yields actually declined in the years following initial development (Elahl and Khushalani 1991). Ironically, much of this farming is small scale, with farms typically less than 8 ha in size and operated by families and small businesses. The capital expenditure for these schemes has been vast, of the order of 320 million United States dollars (USD) for the most recent scheme at Rahad. Many experts, as well as the farmers, question the wisdom of these investments, given the difficulties that have ensued.

The major concentration of irrigated commercial farming activity in sub-Saharan Africa is in the south and east, particularly in South Africa, Zimbabwe, Namibia, and Kenya. Southern African countries have introduced extensive irrigation, primarily from groundwater sources, to an essentially arid and semi-arid environment. Average rainfall in South Africa, for example, ranges from 1 200 mm in coastal KwaZulu–Natal, to 400 mm in the Orange Free State, to 100–200 mm in the northern Cape Province. The average for the country as a whole is 500 mm. Evaporation rates for open water systems in these areas are 1 500–2 500 mm/year (up to 4 000 mm/year in Namibia). Reliance on rainfall alone, therefore, is insufficient to conduct this kind of farming (Davies and Day 1986). As a result, complex, capital-intensive systems of riverbasin irrigation, rainwater dam, and groundwater (borehole) irrigation have been developed. These systems are fairly energy intensive, relying on electrical (in some cases, wind-driven) pumping systems to bring the water from source to end-use. Riverbasin and rainwater dam irrigation is, of course, subject to high losses through evaporation and leakage; groundwater (borehole) irrigation is typically associated with open spraying, which also leads to significant evaporation losses (Davies and Day 1986).

Riverbasin irrigation is widely practiced in South Africa, especially in the Vaal and Orange river systems. Much effort has been devoted to the rationalization of these systems. The latest is the massive Lesotho Highlands Water Project, which will theoretically result in increased efficiency on the supply side through improved storage at source, in the high plateau of Lesotho, where much of the Orange River catchment is located.

The Nile, the major example of riverbasin irrigation in Africa, has been particularly heavily used by Sudan (Adams 1992). However, the people around the Nile still depend largely on traditional floodplain irrigation, rather than on commercially piped or channeled irrigation. Comparable use of river-based irrigation can be found in connection with major dam developments like Akosombo Dam in Ghana, Kossou Dam in Côte d'Ivoire, and Aswan High Dam in Egypt.

In many cases, development of dams has been driven primarily by energy needs, rather than by irrigation needs. Three of the largest hydroelectric developments in Africa, namely, Kariba in Zimbabwe–Zambia, Kafue Gorge in Zambia, and Cabora Bossa in Mozambique, have resulted in little commercial irrigation because of the low population densities and relatively poor agricultural soils in the areas adjacent to the impoundments. There is, however, substantial informal use of waters from Lake Kariba on the Zambian side. Although only about 6 000 ha at present, the area could be expanded four- or fivefold if plans to initiate commercial wheat farming upstream of Victoria Falls are implemented (Kasimona and Mkwaya 1995). The proposed use of the Zambezi's waters above Lake Kariba for municipal, agricultural, and industrial water requirements is now a major area of contention (see "Industrial use").

Traditional or communal agricultural use

Although dominated by commercial farming, agricultural use of water in Africa also includes a large number and range of indigenous or traditional forms of water distribution for cropping:

- seasonal diversions of river flooding along the Nile and in smaller rivers such as the Pongola in northeastern South Africa;
- smaller-scale and less-commercialized floodplain irrigation along rivers like the Hadejia and Jama in Nigeria and the Zambezi in the Lozi tribal areas of Zambia; and
- small-scale irrigation systems based on river flow or on small rainwater dams in many countries, including Nigeria, Madagascar, Tanzania, Senegal, Mali, Sierra Leone, Sudan, Burundi, Chad, and Somalia.

Although small-scale irrigation is usually thought to be minor in extent and impact, it actually constitutes almost 50% of irrigation in Africa by land area, or 2.38×10^6 ha out of an estimated total of 5.02×10^6 ha (FAO 1986). Such forms of irrigation are indigenous in two senses: many predate colonialism, and most are locally controlled.

Adams (1992) argued that these traditional systems of water catchment and distribution are typically as efficient as and commonly more conservative than large-scale systems and therefore deserve far more attention from planners and financial agencies. Whether this is so depends on a range of variables: the maintenance provided to the system, the types of crops being irrigated, and the methods used to carry water to the crops. Adams' use of *efficiency* is probably closest to what Xie et al. (1993, p. 5) called "conveyance efficiency," that is, efficiency in carrying water from the source to the distribution system. Whether such traditional systems are inherently more efficient in terms of the local provision and end-use of water (or water *distribution* and *application*, to use the terms of Xie et al.) is a moot point, especially in view of the controversy over the extent to which irrigation is a consumptive use of water (Seckler 1993).

Municipal use

Municipal uses of water include supply to the domestic sector, as well as to commercial buildings and to washing facilities. Municipal use represents the second largest use (though a distant second) in Africa, at 7% (Xie et al. 1993). However, this use ranges from 81% in Equatorial Guinea, 72% in Gabon, and 63% in Zambia to less than 1% in Madagascar and Sudan (Xie et al. 1993). Generally speaking, the higher percentages are probably a function of conditions (such as war in Mozambique or decommercialization of agriculture in Zambia) that have reduced agriculture's dominant role. The lowest percentages, on the other hand, are a function of the very dominant role of irrigation agriculture in countries such as Sudan.

Urban water-reticulation systems in most African cities were originally developed by colonial governments and sized to meet the needs of a much smaller population. Extending water supply to rapidly expanding urban and peri-urban communities populated largely by rural migrants and workers has proven a daunting challenge. As an example, domestic water demand in South Africa is expected to grow by about 2% per year over the next 30 years: if this rate of growth is realized, demand in 2020 would be three times that of 1970. This would also increase the share for domestic users: by 2010, more than 17% of South African water use would be domestic, compared with 9.3% now (South African Department of Water Affairs 1986). This trend will, if anything, be strengthened by the recent decision of the new government to provide substantial low-interest loans and subsidies for low-income housing, which will inevitably result in a substantial increase in serviced lots and, correspondingly, in requirements for centrally supplied water.

As a rule, municipal water in African cities, as in European and North American ones, is supplied from open reservoirs fed by dammed rivers or by direct rainwater accumulation. In arid and semi-arid areas, there are also numerous examples of groundwater- or borehole-based municipal supplies, although these usually augment reservoir supplies. The most striking example of the use of boreholes to supply a major city is Windhoek in Namibia, which has also recycled sewage for many years as a water-conservation measure.

Municipal water supply in Africa illustrates perfectly the dilemma created when there is competition for use of an underpriced resource. Few African countries have been able to deal with the rising demand for safe, reliable municipal water supplies because they have already allocated much of the river-based and groundwater supplies in surrounding regions to agriculture and, in a few cases, to industry. Because these bulk uses are often underpriced relative to the long-run costs of system development or even to the short-run costs of operation, governments do not recover sufficient revenue to pay for the very large investments required for expanding domestic and commercial-sector reticulation systems. As a result, several countries are experiencing substantial urban water-supply crises.

The problem of domestic water supply in rural areas is entirely different. In West Africa, there are numerous examples of rural domestic water being supplied from irrigation systems (rivers or lakes). In South, Central, and East Africa, however, most rural villages depend either on boreholes or, much more commonly, on water from nearby rivers or lakes. Tanzania is an exception: a number of small, traditional irrigation systems in areas such as Kilimanjaro and Lake Natron supply both domestic and agricultural needs (Adams 1992).

Industrial use

Industrial use of water is still a relatively minor factor in Africa. Estimates place total African industrial use at 5%, the lowest among the six major regional groupings of nations. However, this small average masks substantial variation, from 25% in Zaire and 22% in Lesotho to virtually nil in Sudan, Somalia, Burundi, and Madagascar. Further, industrial use is likely to grow faster than use in other sectors as African countries undergo major economic changes over the next 10 years. Because industrial water use is generally more efficient than agricultural use, this shift may presage an overall improvement in efficiency, but it is unlikely to signify a reduction in overall water use (Adams 1992).

By far the most significant demand for water from the African industrial sector comes from mining and metallurgy (smelting and refining). Coal mining in Mozambique has been estimated to require up to 4 m^3/t or 1 m^3/s of water in the mining and washing processes combined (David 1988). Large diamond mines, such as that at Orapa in Botswana, and aluminum smelters, such as that at Richards Bay in South Africa, use substantial volumes of groundwater in the processing of minerals. Use of water at Orapa is large enough to reduce water levels at the major source of this groundwater, the Okavanga Delta.

In some countries, like Zimbabwe, Kenya, Botswana, Namibia, Tanzania, and South Africa, water is used to generate steam, both for electricity and for industry, and to cool coal- or oil-fired generators. Inasmuch as many generators are located in semi-arid areas, most of this requirement comes from groundwater. The South Africans have, however, pioneered in the use of air-cooled generators and have succeeded in substantially reducing demand on local groundwater supplies in arid areas such as the eastern Transvaal (Davies and Day 1986).

Hydroelectric power constitutes a small but important end-use in Africa. A number of major riverbasin developments are tied to hydroelectric dams, including those on the Volta, Zaire (at Inga), Orange, Ruacana, Kafue, Zambezi, and Nile rivers. Several of these projects resulted in substantial back-flooding and reservoir formation and simultaneously reduced the impact on agriculture of traditional seasonal flooding downstream of the dam. Yet, in general, it may be said that the use of water for hydroelectric power is still in its infancy in Africa. The World Bank estimated that new projects, in various stages of planning, could collectively increase hydropower's contribution to Africa's energy supplies by a factor of seven, assuming there is increased integration of power transmission systems to justify the increased investment in hydroelectric generating capacity. Although these figures are probably excessively optimistic and may underestimate the cost of delivering electricity from remote sites in politically unstable countries such as Zaire, it is clear that the use of riverbasins for hydroelectric power will increase substantially in sub-Saharan Africa over the next 25 years.

Conclusions

Because water use in Africa is so heavily dominated by agriculture, it is appropriate that concerns over demand management focus on this sector. Opportunities for reducing agricultural demand are considerable, particularly for large-scale commercial farms, and agricultural use of water on such farms is extremely responsive to price.

On the other hand, agriculture's pivotal role in many African economies also makes it a target of political strategy. This is particularly true for traditional or communal agriculture. African governments, assisted in many cases by donors and multilateral banks, have spent large amounts of scarce capital on dam and irrigation projects that are essentially supply-side initiatives. By world standards, the costs of these initiatives are extremely high, ranging from 1 000 USD/ha for smaller, locally funded projects to 8 000 – 20 000 USD/ha of reservoir surface for large-scale capital projects (Wyss 1991). Although these initiatives have done little to alleviate water shortages and, in some cases, may actually have worsened them, it seems unlikely that governments would voluntarily place the burden of reducing demand on the user, rather than simply seeking more support for supply, no matter how promising the prospects for correcting the water crisis. Pressure for improvement in DSM in agricultural use must come from outside: from local pressure groups and donors.

The situation for industrial and domestic uses is more positive. These sectors continue to have a very small impact on African water balances, although their role is growing steadily. Indeed, in a few countries, notably South Africa, Namibia, Zimbabwe, Sudan, Ethiopia, and Egypt, the crisis point has already been reached in the larger urban areas. This crisis has been brought about by a combination of increased industrial uses and the pressure from burgeoning urban and peri-urban townships for water access. Bulawayo, Zimbabwe's second city, is an example. The city's continuing water shortage over the past several years has severely affected both domestic and industrial use and has lately produced a shift of population and industry away from the town. Solutions to the crisis centre almost exclusively on water-supply megaprojects, and virtually no efforts have been made to deal with the question of demand, except through emergency quotas and penalties for overuse.

Nevertheless, crises of this sort have had a salutary effect on planning in some countries: the crises bring water to the forefront of public debate and help focus discussion on water needs, or at least on perceived needs. Though these crises may be solved by patchwork supply increments, there is some hope that realization of the immense costs of long-term supply-based solutions will force reexamination of the need for demand-based solutions.

Africa's water-management crisis

Africa's water-management crisis is well documented. At least 10 African countries, virtually all in North Africa and the Sahel, were officially classified by the World Resources Institute (1992) as "water-stressed." It is expected that at least six more will be added to this list by 2025 (Bryant 1994). Moreover, these figures exclude South Africa, where the sources of stress are both overuse and environmental limitations, and those other countries of southern and Central Africa that have recently experienced severe drought conditions (Zimbabwe, Zambia, Botswana, Mozambique, Malawi, and Namibia) and are usually classified as semi-arid. Taken together, well over half of the countries in Africa have significant water-access problems stemming from a combination of overuse, poor management, and adverse climatic conditions. Indeed, some authors argue that per capita water supplies in Africa have actually declined by as much as 50% since 1950 (Bryant 1994).

Much of this crisis centres on growing demand for agricultural water. Most African countries depend heavily on agriculture as a source of income for rural populations, and many have also developed substantial export markets for products such as tea, coffee, sugar, cocoa, rubber, tobacco, and flowers — crops that require reliable and abundant water supplies and earn substantial foreign currency.

The crisis of African water management cannot, therefore, be solved simply by reallocation of uses among sectors, that is, from agricultural to domestic or industrial uses. Nor can it be solved by increasing supplies: in most water-stressed countries, the major readily accessible sources of water supply have long been identified and, for the most part, developed. There are plans to develop a few others soon. The sole remaining solution is to limit demand without seriously degrading the quality of life of Africa's burgeoning populations. The prognosis is not good. As Davies and Day (1986, p. 120) pessimistically argued for South Africa, failure to deal urgently with the issue of water demand and efficiency could have dire consequences:

> It is generally assumed that the 'best case' we can realistically expect, a combination of the greatest possible use of surface resources, the slowest population growth and negligible exploitation, will bring the drought on for good in 2020.

The experience of developed countries versus that of developing countries

Searching for solutions to this crisis inevitably brings us to compare the recent experience of developed countries with that of developing countries. However, the comparison is not necessarily a helpful or instructive one: developed countries have mismanaged their water resources just as much as developing ones, and in many arid areas of the developed world, the water-supply situation has also reached a crisis point.

The advantages that developed countries have over developing ones are, of course, their stronger revenue base and better access to efficient technologies. Most western countries have well-developed agricultural and domestic water-supply systems supported by a combination of user charges and general tax revenues. Cost-reflective pricing is still a rarity, particularly for agricultural users, although it is attracting increasing attention. As Frederick (1993, p. 22) showed for the member countries of the Organization for Economic Co-operation and Development, "markets and prices have rarely been the principal or even important mechanisms for allocating water resources." In North America, agricultural (irrigation) tariffs remain well below cost-recovery levels, despite major conservation efforts.

Several studies have identified the need for a demand-based approach to water management in the industrial countries. These studies describe numerous efforts, particularly by the United States, to increase the efficiency of water delivery and use. For Canada, Brooks and Peters (1988) suggested a water-demand management strategy that would first identify sectors where water supply was becoming a problem, then try to remedy the barriers to demand management, and finally derive a sector-by-sector program combining regulatory and pricing measures with marketing and technology-transfer measures. Their study showed that the identification of barriers to demand management is particularly important:

- Economic barriers involve the undervaluing of water by consumers, underpricing by suppliers, and failure to charge self-supplied users for water rights.
- Market failures include lack of adequate information on conservation opportunities, lack of access to affordable capital, unnecessarily restrictive investment criteria, general indifference to the cost of water, and separation of the costs of conservation from its benefits.
- Organizational barriers are characterized by a preoccupation of authorities with supply and with engineering solutions and by concern over the financial risks of demand management.

These barriers are virtually universal and could be applied with minor adjustments to developing countries in Africa. However, there is an overriding difference between experiences in northern industrial countries and those of countries in Africa. The former countries are involved in minor adjustments to relatively prosperous market economies in which there is an established pattern of treating most resources as commodities. The latter countries, on the other hand, are in various stages of transition to a market economy in which resources, traditionally regarded as public or communal, are only beginning to be treated as commodities. In the imperfect and transitional market economies of many African countries, the economic and market barriers are truly substantial and diverse. Nevertheless, many of the same problems exist: lack of information, lack of capital, low or indifferent pricing, traditions of usufruct, and high financial risks.

The experience of Africa

The African continent has generated a number of significant endeavours in water-demand management, although the bulk of effort has been concentrated on the improvement of conveyance, rather than on distribution or end-use efficiencies.

The following sections discuss four areas where demand management either has been practiced in the past or is being considered for the future:

- management of riverbasins;
- management of municipal water supplies;
- management of irrigation sytems; and
- use of pricing as a demand-management tool.

MANAGEMENT OF RIVERBASINS — Riverbasins should be a major target of demand-management activities in Africa. As Adams (1992) noted, virtually every river in Africa has now been dammed for either agricultural (irrigation) uses or electricity generation or, more often, both. The Zambezi, the Kafue, the Senegal, the Nile, the Vaal, and the Niger and its tributaries have multiple dams, and there are plans for more. From a planning perspective, Africa has a surprising number of cases where riverbasins affect more than one nation. Of the 13 cases in the world where five or more nations form part of a river or river–lake basin, Africa has 5: the Niger, Nile, Zaire, and Zambezi rivers and Lake Chad (Gleick 1989).

As a general rule, damming of major rivers in Africa has taken place with minimal international (or regional) cooperation and, hence, with little attention to either downstream or upstream impacts. Nevertheless, although demand management has mostly been neglected, sevearal efforts have been made to reverse this historic tendency and institute multinational planning systems to improve the efficiency and minimize the negative impacts of water use in riverbasins. Examples are the Zambezi Action Plan (ZACPLAN) in the Zambezi basin area, the Volta River Authority, the Niger Development Board, and the Nile River Water Agreements.

The consolidation of water-supply management in riverbasin authorities naturally involves a political element, just as it does with agricultural water uses in Africa. As Scudder (1989) pointed out, economics invariably takes second place to political considerations in riverbasin projects, with the result that supply concerns tend to take precedence over demand concerns. The ZACPLAN omission of demand considerations perfectly illustrates this: neither the donors nor the governments involved in this ambitious scheme wish to consider the intractable problem of whether the demands for Zambezi water use are legitimate and have been accurately quantified, and neither the donors nor the governments have considerated mechanisms for limiting this demand.

MANAGEMENT OF MUNICIPAL WATER SUPPLIES — In some respects, domestic water supplies in Africa represent a more tangible crisis than is found in any other sector. The reason for this is simple: although agricultural and industrial uses may be under more objective threat, scarcity of water for households affects the population, particularly the middle-income population living in cities, more directly. Disproportionate attention is therefore given to the water crisis in municipal areas.

Solutions to the problem of domestic water supply in arid and semi-arid or otherwise water-stressed areas have been many and varied. On the supply side, the solutions include the following:

- expansion of conventional supplies through increased damming of rivers and streams;
- development of boreholes on a large scale, typically in combination with damming; and
- interbasin transfers.

Supply-side solutions of this sort are still being considered and actively planned in many African urban centres. The burgeoning peri-urban populations resulting from rural migration have created a need for substantial new supplies. This, in turn, has prompted urban planners to look for new sources. For example, Bulawayo in Zimbabwe has already utilized all of the available river-based sources; six major dams serve the 350 000 residents of this city. The alternative, using boreholes drawn from aquifers, has been exploited as an emergency option, but this is not sufficient to meet the expected needs of the city. As a result, effort has shifted to megaprojects, such as the proposed Zambezi–Bulawayo pipeline.

On the demand side, solutions include the following:

- establishing seasonal quotas and usage restrictions;
- rationing by time of day or area;
- recycling domestic and, if feasible, industrial wastewater, including sewage; and
- setting price penalties.

These actions apply primarily to urban water use. Water use in rural areas outside organized municipal authorities is, of course, less, both per capita and absolutely, and is less likely to be subjected to restrictions, although many rural areas in Africa are certainly water stressed.

Some of these solutions are practiced to a certain degree. For example, the development of quotas is restricted primarily to emergency situations, such as those experienced by municipal suppliers in relatively developed cities like Johannesburg, Harare, and Nairobi. Time-of-day rationing also occurs, primarily in emergencies similar to that caused by the 1991–92 drought in Central and southern Africa. However, such rationing is becoming fairly commonplace in many cities in areas of the continent that have seasonal rain: because local reservoirs and dams are only rarely recharged to their full extent, they require substantial restrictions on use toward the end of the dry season. Very few examples of price penalties are found in Africa.

MANAGEMENT OF IRRIGATION SYSTEMS — Efforts to manage African irrigation systems for greater efficiency are numerous in both large-scale commercial and small-scale traditional forms. Recent reviews of this subject by the World Bank show conclusively that management of large-scale commercial irrigation schemes is problematic in much of sub-Saharan Africa. On the whole, small- and medium-scale schemes that are locally managed and employ traditional methods tend to fare better in terms of both economic efficiency (that is, rate of return) and water-use efficiency. Lele and Subramanian (1991, p. 60), for example, suggested that, in contrast to large-scale schemes in Africa,

> small-scale irrigation has had largely positive experiences and this offers considerable untapped potential for expansion involving both public and private investment. . . . Where the private sector is involved, an effort should be made to decentralize irrigation systems and make their clients directly accountable for them.

The difficulty with this argument, as the authors themselves admit, is that few governments have been willing to forego investment in large-scale schemes simply because they are inefficient. Large-scale systems are more amenable to central-government control and, therefore, more attractive to the entrenched urban bureaucracies of many African states. Moreover, involvement of the private sector is not really a pertinent issue in many countries. Small, locally managed irrigation systems are more often "public" than "private" in terms of effective ownership and management philosophy. Nonetheless, the localization of control still confers significant efficiency benefits. It is important not to conclude, as the World Bank has done in several studies, that private ownership of the resource or private-sector management of its distribution will necessarily bring about increased efficiency (Wyss 1991).

Adams (1992) presented a very cogent argument for the inherent efficiency of small-scale, traditional irrigation systems. He showed, for example, that small-scale furrow irrigation in Tanzania by the Chagga of the Kilimanjaro region and by the Sonjo of Lake Natron is very carefully managed and that, because of this, the Tanzanian government has been able to use traditional irrigation as a basis for further technical improvements, rather than replacing it with modern large-scale systems. Similar examples of productive traditional irrigation systems may be found in the rainwater-harvesting techniques of northern Kenya, the use of terraces for erosion control in Burkina Faso, and the Molapo Development Project of the Okavango Delta in Botswana (Adams 1992). Adams cautioned that intervention in such indigenous or traditional irrigation and water-management systems has drawbacks: for example, altered perceptions of ownership and responsibility, increased economic uncertainty for smallholders, and increased dependence on outside support and technology. The solution to this is not to leave indigenous systems alone on the premise that they work better without interference but, rather, to insist that "integrated resource management" be followed and that information about new technologies and training in their use be provided as part of the development package (Adams 1992).

Much the same argument could be applied, of course, to virtually all aspects of irrigation development and rehabilitation. Locally oriented solutions are preferable because of the high cost and economic risk associated with externally imposed solutions and because of the "hidden hand" of environmental damage and possible loss of user control over water management. On the other hand, uncritical acceptance of local solutions is little better: a large number of small-scale irrigation systems can be just as environmentally damaging in the aggregate as a few large-scale ones and may be less efficient in their application than they could be with appropriately scaled technological improvements and proper information and training.

The private–public argument deserves closer scrutiny. In southern Africa, particularly, the confrontation between private (commercial) and public (subsistence) agriculture has been a major political and economic management issue. Because most of the countries of this region are arid or semi-arid, the issues of access to and management of water resources are a central part of the debate. Commercial agriculture has thrived in countries like Zimbabwe, Namibia, and South Africa in part because of substantial public investment in irrigation infrastructure and support systems. This investment includes government-funded programs for damming of major rivers, provision of free technical advice and water monitoring, and subsidized loans for boreholes and private dams. Indirect support has been provided in the form of price subsidies to farmers for the products of irrigated agriculture, along with price controls on the raw materials they require to develop this agriculture. The result has been an economy with built-in distortions favouring commercial (mainly white, male) farmers over communal or traditional (mainly black, female) ones. The latter do receive some of the benefits of agricultural extension work, but investment, particularly in irrigation, has been minimal, except recently in some communal areas of Zimbabwe and Zambia (Kasimona and Makwaya 1995).

The pertinent issue for this paper is whether disproportionate investment in the modern or commercial sector has increased or reduced the efficiency of agricultural water use in the countries concerned. The answer is not simple. For example, the commercial sector's use of large-scale spray irrigation for tobacco, maize, wheat, and other cash crops has resulted in higher evaporative losses than occur with, say, drip irrigation. On the other hand, evaporative losses are low compared with losses from flood irrigation or even canal irrigation. In addition, some field crops, such as wheat, cannot be effectively irrigated by any other method. Still, the investment and operational costs of spray irrigation are very high, and the successful operation of these systems requires careful and informed management. Davies and Day (1986), for example, argued that efficiency levels for spray irrigation are of the order of 30% and that the low efficiency of these systems is compounded in South Africa by poor maintenance, erosion damage, and salination effects. Others have shown that the on-farm efficiency of spray irrigation in Israel and other parts of the Middle East can be as high as 75–80% and in hot, dry areas with maximum evaporation loss can be as high as 60% (Schwarz 1991).

As these discussions suggest, investment in spray irrigation, drip irrigation, or improvements to irrigation-water distribution (for example, by lining canals) are capital intensive. They almost always require some degree of public investment, whether the eventual beneficiary is a private commercial farmer or an indigenous communal farmer. In southern Africa, governments have supported major public-investment campaigns to help the commercial farming sector be competitive internationally. These governments have invested in irrigation and other water-supply measures as a central part of this economic philosophy. Correspondingly, these governments either have failed to invest or have invested on a much smaller scale in water-management schemes to support the communal farmer, and political confrontation, economic mismanagement, and adverse ecological impacts have ensued.

In the Save River catchment area of southeastern Zimbabwe, for example, traditional dependence on floodwater irrigation, coupled with massive government-induced increases in the local indigenous (smallholder) farmer population, has resulted in severe degradation of the entire river system. The river now flows (when it flows at all) well below the soil surface and is effectively inaccessible. In northeastern Zimbabwe, near the Nyanga Mountains, development of large irrigation dams has pushed traditional farmers (for whom the irrigation was intended in the first place) into marginal areas because there is insufficient cultivable area near the new dams. Nevertheless, small-scale commercial farming, using a combination of groundwater (boreholes), summer-flow storage dams, and off-river flow for irrigation, has been relatively successful and efficient in Zimbabwe, as documented by ter Vrugt (1990).

Other examples of efficient irrigation based either on technological improvements or on better management and operational control can be found in Africa. Xie et al. (1993) summarized the essential factors in efficient management:

- ensuring a reliable water supply to the end-use point;
- assessing soil characteristics and plant requirements;
- improving management skills;
- improving maintenance;
- using surface and groundwater conjunctively;
- disseminating information on efficient technologies and techniques; and
- implementing demand management through legislative and administrative intervention.

To this list, Grainger (1990), who has studied the effect of improved water management on the desertification process, would add four items:

- improving project design to ensure that drainage is provided for and farmer needs are being met by the design supply;
- expanding farmer involvement in the design and implementation phases;
- promoting more attention to rehabilitation over new construction; and
- increasing use of small-scale approaches that are within the competency and scale of small farmers and that are effectively decentralized in development and management.

USE OF PRICING AS A DEMAND-MANAGEMENT TOOL — The use of pricing as a management tool is relatively underdeveloped in Africa. Although most irrigation systems pass the cost of water on to the consumer in some form or other, as do most domestic and industrial water-provision systems, the tendency is to underprice this resource or to use price to control access but not the level of actual use. Correct pricing of water resources, therefore, presents a significant opportunity for DSM.

The most common pricing strategy in water management, as in energy management, is to cross-subsidize among various consumer categories. As a general rule, such subsidies take the form of overpricing domestic water use on the premise that domestic users can more easily conserve than agricultural or industrial consumers and are more likely to be able to afford such an increase without serious stress. The equity implications of this could be considerable, but, to the extent that low-income domestic consumers get their water from central water sources (communal pumps and pipe outlets) and, therefore, do not pay directly for the resource, adverse equity effects are mitigated.

In Zimbabwe, agricultural water use is effectively subsidized through a "blended price," which ameliorates the cost to the consumer of new irrigation developments. Although blending ensures that new entrants to the market do not bear an unreasonable cost burden, it tends to depreciate the value of water and thus reduces interest in water conservation (Helming 1993). This is a cross-subsidy in the sense that other customer classes are paying directly or indirectly for the underpricing of water to farmers.

Experiences from elsewhere in Africa suggest a similar picture. The true cost of irrigation development is rarely passed on to the farmer, partly, at least, because so many irrigation projects are funded by donors. Where the project is based on loans rather than grants, as in World Bank projects, there is a tendency to enforce cost recovery as a matter of principle, and requirements of this sort are often embodied in loan covenants. However, such cost-reflective pricing will only succeed where the producers are able to recover fair prices for the sale of resultant crops. The price increase to water users must therefore be balanced by a removal of price-setting on food intended to protect consumers (Barghouti and Subramanian 1991).

Hydroelectric generation is a significant element in a number of African riverbasin schemes, and, indeed, prospective revenue from electricity sales is a major factor in decisions to invest in such schemes in the first place. Commonly, however, the water used for generating hydroelectricity is regulated by one organization but a different organization is responsible for, and derives revenue from, electricity generation. Generating electricity from water represents an opportunity cost (other uses of that water, such as for industry, recreation, or irrigation, having been reduced), so it is appropriate to charge a rent to the generating authority for the use of this resource. Ideally, this would create a revenue stream for the water authority that could be used for further development and maintenance of the hydrological system, provided, of course, the money is not diverted to general revenue by government! In addition to improving economic efficiency, charging a rent for the resource also provides an incentive to conserve, inasmuch as the utility is charged only for water actually used in generation.

A study is under way on the Zambezi River system to develop just such a tariff to fund the Zambezi River Authority (ZRA). If approved, the tariff would be paid to ZRA by the national utility of each country — the Zimbabwe Electricity Supply Authority (ZESA) and the Zambia Electricity Supply Company (ZESCO) — on the basis of electricity actually generated from the Kariba and Victoria Falls power stations each month. This would replace the current system whereby the two utilities are given an annual allocation of water based on their theoretical requirements but no rent is levied and there is no penalty for exceeding the allocation (P. Robinson, Harare, Zimbabwe, personal communication, 1994).

Many similar opportunities to use pricing as a mechanism for water-demand management exist in Africa. However, pricing is rarely a solution in itself. For example, most small-scale farmers have not had any significant influence on the type of irrigation or other water-delivery systems they use. Thus, raising prices to these farmers, who may be using an inherently inefficient or inappropriate delivery system, will not in itself improve conservation of water and may, in fact, lead to economic distress. On the other hand, Africa is full of examples where authorities failed to provide sufficient revenue to pay for the maintenance and periodic renovation of irrigation systems, with the result that the systems quickly fall into disrepair, with attendant losses in efficiency. It is clear, therefore, that, for irrigation agriculture to operate efficiently, there must be some form of cost recovery, whether through the pricing mechanism itself or by some other means.

Pricing reform means effectively that water is treated as a commodity rather than as a communal good or right. Although this argument has more merit in the context of market economies and large-scale commercial projects, where there are substantial investment and operational costs to be recovered, it can also be applied to small-scale, traditional systems. In Kenya, for example, the efforts to augment and rationalize small-scale irrigation systems by making them the cornerstone of market agriculture are a case in point. Water rights, though traditionally supported, are now seen as something that must be earned through proper maintenance and efficient operation of the system (Adams 1992).

In much of Africa, the driving force behind the shift toward market-based approaches is the economic-reform program initiated by the International Monetary Fund or the World Bank and supported by many bilateral donors. Commercialization or privatization of government and parastatal organizations has become commonplace, if not always successful. Water providers (riverbasin, irrigation, and municipal water authorities) have largely fallen outside this initiative, possibly because they are traditionally dominated by public interests in the industrial countries as well. However, there is pressure for water authorities to operate on a profitable basis to ensure that they have sufficient revenue for ongoing management and operational costs. This, in turn, calls for a move toward cost-reflective tariffs for water, just as it does for energy. According to the World Bank (1992), the real costs of water have typically doubled for every new irrigation or municipal water project in Africa, by itself a strong argument for continual upward adjustment of prices.

Demand-side management for water: a basic paradigm

Defining demand management and demand-side management

As applied to water resources in particular, *demand management* is a term that usually refers to a broad range of techniques and processes based on the end-use requirements of water consumers, rather than on the supply requirements of water providers. *Demand-side management*, by contrast, is a term used to describe a somewhat narrower set of activities and principles that are typically initiated by the resource provider (utility) itself as a part of its corporate-planning and capital-investment process. In the energy context, DSM is closely associated with efforts to alter utility-load patterns. DSM is also associated with specific kinds of planning, often referred to as "integrated resource planning" or "least-cost planning," in which demand reduction and supply increase are given equal weight by a utility when making investment choices.

By definition, DSM requires a better understanding of consumer requirements. As a result, power utilities implementing DSM are more responsive and more likely to make socially appropriate investment decisions than would traditional supply-oriented utilities. DSM allows an "integrated look at technology options, customers' needs and utility considerations" and recognizes that "customer needs" (demand) are not fixed and can be manipulated by both economic and pricing incentives (Gellings and McMenamin 1993, p. 1).

In its most conservative form, DSM refers only to the utility's efforts to achieve a specific "load shape" that allows it to meet capacity needs at minimal cost. In this definition, the "time pattern and magnitude of the utility's load" are the only criteria (Gellings and McMenamin 1993, p. 2). However, the application of DSM also includes activities such as customer-initiated efficiency investments, customer generation, promotion of new applications for electricity, information programs intended to bring about strategic conservation in different sectors, and a variety of marketing strategies designed to influence the share that electricity holds of the energy "pie" (Gellings and McMenamin 1993).

The most significant implication of DSM is its impact on resource planning. The convention in energy, as in water, has been to estimate demand based on historical data, using established elasticities of income and price to predict the rate at which demand will increase or, in some cases, employing simple trend analysis. This practice invariably results in generous assumptions about demand growth; supply requirements are chronically overestimated, as has been the case, for example, at Ontario Hydro, Canada's largest utility, which now has a substantial overcapacity and has been forced to close all or part of several generating stations.

By contrast, DSM is characterized by planning that takes account of all possible investments on both the demand side and the supply side (integrated resource planning). As a result, DSM planners are able to use least-cost planning criteria, selecting investments on a more objective basis, regardless of whether they entail new generation capacity, additional transmission capacity, or consumer incentives to purchase efficient end-use technologies.

Advantages and disadvantages of demand-side management

DSM programs have, for the most part, been implemented by utilities in various industrial countries. There are several reasons for this:

- Power supplies in the industrial countries are run on tight margins, either by private power suppliers or by public power suppliers operating on commercial criteria. Investment in new supplies is therefore closely scrutinized, and decisions must reflect the opportunity cost of foregoing alternative investments.
- The price of electricity in most industrial countries is relatively high and generally approximates the long-run marginal cost of new supply options, which makes efficiency investments more attractive than they would be in low-price regimes.
- Customers in industrial countries are sensitized to price and react quickly and fairly predictably to price increases. They also have the capacity to deal with price increases by selectively reducing demand (for example, by purchasing more efficient appliances or other end-use technologies). Their sensitivity and responsiveness are, of course, partly a function of higher income levels and more sophisticated consumer markets.

Utilities initiating DSM in the industrial countries usually have a fairly precise notion of the supply cost of conservation options, that is, the discounted cost of an investment in efficient end-use technologies for a particular customer or class of customers. Comparing this with the discounted cost of supplying the same amount of energy provides a more realistic planning base. A utility typically does this by developing conservation–supply curves, which compare the cost per kilowatt–hour saved for different conservation options. These can then be objectively compared with the cost per kilowatt–hour of different supply options. For example, the conservation–supply cost to the utility of a technology such as compact fluorescent lights might be 2.5 cents/kW · h of energy saved over the life cycle of the product, against a delivered cost of 4–5 cents/kW · h of energy supplied from a conventional power station. If the utility can mount a promotional campaign or offer a "prescriptive" subsidy on the purchase of such products at a total cost that is less than the avoided cost of the conventional station or other supply options, then this represents a realistic alternative investment, one on which the utility could earn an attractive rate of return. Even though this investment might only defer, rather than displace, the supply option, it can still produce substantial savings to the utility by delaying the need to embark on expensive capital investments.

DSM has become a major part of utility planning in many industrial countries and provides a distinctively customer-oriented focus at a time when many utilities are seen by their customers as frankly predatory and profiteering. However, success has proven difficult to reproduce in developing countries, particularly in Africa, where the provision of electricity is typically a state monopoly and where commercial criteria have, for the most part, not been successfully applied. Several important caveats apply to the use of DSM in developing countries:

- DSM does not work as well where electricity prices are low. Because of political factors, electricity prices in many jurisdictions in Africa have been artificially depressed, although this is changing rapidly. It is, of course, possible to use nonprice incentives, such as information programs, where pricing is still too low to encourage voluntary consumer investment.

- There must be a scarcity of the key resource (energy or water) to compel the national utility to pursue some form of investment strategy aimed at coping with increased demand. Where energy (or water resources) are available and well distributed, there is far less incentive to undertake DSM.

- There must be a substantial investment by the utility in both new planning methodologies and customer-information databases. Hence, the utility must maintain a good cash flow and strong revenue base, which are major obstacles among African utilities at the moment.

- Barriers to acquisition of efficient technologies, including tariff or import barriers, must be removed or minimized to ensure that these technologies can be acquired at a fair and competitive price. Protectionist policies of the sort found in many African states until recently are a disincentive to implementation of consumer-driven DSM.

- Consumers must have access to good market and product information without excessive cost. This implies a strong and well-managed information program and a sound statistical base. These are again difficult to achieve in many African utilities, which often lack the revenue base and human resources to carry out such programs.

Notwithstanding these significant limitations, a number of African utilities have already embarked on DSM programs, and others are closely studying the prospect for DSM. Utilities in relatively wealthy countries, such as ESKOM in South Africa, are, of course, better positioned for this kind of initiative. ESKOM has undertaken fairly detailed assessments of DSM prospects and has, in fact, initiated a limited "small customer" DSM program, despite the fact that it has a 25% overcapacity! Poorer utilities, such as Electricidade de Moçambique in Mozambique, ZESCO, and Tanzania Electricity Supply Company, have also embarked on such programs (with donor assistance) because they are faced with significant supply shortages and they badly need to encourage consumers to use electricity more efficiently.

Conditions for success in demand-side water management

By reviewing the pros and cons of energy-DSM programs, one may deduce several conditions for successful water-DSM programs:

- Detailed knowledge of end-uses, or customer consumption patterns, is a precondition of successful DSM programming, as suppliers need to be able to target specific customer classes for incentive or subsidy. Electricity utilities meter consumption for most classes of user, so the basis for a detailed consumer (end-use) database already exists. However, this is not so for water utilities, although the gross requirements of different classes of consumer are generally known and many specific efforts to measure usage have been undertaken. Metering of flows to irrigators, measurement of water passing through and consumed by hydroelectric generators, and (although rare in Africa) metering of domestic and industrial consumers to provide revenue to municipalities are some examples.

- It is necessary to alter official perceptions of the relationship between water pricing and consumption, particularly those that characterize centralized water-management schemes. Water authorities must establish systems to measure water consumption and then charge according to consumption. In fact, it can be argued that the principle of cost-reflective pricing of water should be attached, not just to river- or rainfall-based irrigation or municipal water supplies, but to groundwater (borehole) supply as well, because use of this sort involves depletion of a national resource for the benefit of a few.

- Pricing regimes must reflect costs. In other words, the overall cost of operating, maintaining, and (perhaps) expanding the water-supply system should be included in the average consumer price charged for water and (possibly) in the prices charged to specific end-users or classes. Such short-run marginal-cost pricing, which accurately reflects operating and maintenance costs but not expansion costs, would be superior to the present system of politically expedient pricing and would be easier to develop and administer than long-run marginal-cost pricing.
- Water suppliers must have general knowledge of the types of consumer end-uses and their typical requirements, namely, the amount of water required for specific kinds of activity, such as industrial cooling, production of beverages, washing, or household consumption. This knowledge will enable the suppliers to identify areas of wastage or inefficiency more readily and to gear their DSM programs to the most wasteful consumers.
- There must be clear lines of authority over the storage, catchment, and eventual distribution of water. Authorities, whether central or local, must be accountable to their consumers.
- There must be access to water-efficient technologies or processes. In practice, therefore, these technologies or processes must be available when and where needed and at a competitive cost.

Implementing demand-side management for water

Analyzing customer end-uses

The first and most obvious task in implementing a successful DSM program for water is to analyze customer end-uses. Doing this with accuracy requires either (1) a system of data collection that is built into the revenue collection or administration system or (2) an accurate means of differentiating customer classes and estimating the number of customers in each.

The first requirement presupposes an organized and centralized water-distribution system, such as we might find in large-scale irrigation schemes in Sudan, or a municipal water-supply system, such as we might find in many southern Africa cities. These systems can fairly easily produce data on average consumption per customer over a stipulated period. Such data, however, are useful only to indicate broad shifts in consumption patterns.

The second requirement assumes that there is a clear division by customer type in the distribution system, perhaps a geographical or locational breakdown, so that different domestic areas can be distinguished on the basis of gross consumption and distinguished again from predominantly industrial areas. Although less precise, an analysis based on this kind of information can be done with minimal skills and can identify average consumption per unit by customer class.

Neither method will provide a detailed end-use breakdown. For example, they will not differentiate hot-water use from cold-water use or, in the case of industry, water for boiler makeup from water used in the product itself. In the absence of metering, achieving this would require an end-use survey: a sample survey of water users to identify typical end-use breakdowns in different subsectors and to permit extrapolation to the population as a whole. This is the method favoured by energy planners and, increasingly, by energy utilities. However, it is an expensive method and is likely justified only for large industrial consumers and for a small sample of domestic consumers of different income levels.

Once an end-use breakdown is obtained, it is possible to generate an analysis of water-quality requirements in each sector and subsector. In the case of industrial water use, for example, this would identify which uses are consumptive and which are nonconsumptive, what the requirement is for pretreated water, whether recycled water from certain internal processes is acceptable, and whether there is a need for (or tolerance of) water from thermal sources (such as recycled cooling water from a thermal generator). In the case of agriculture, an end-use breakdown would determine the specific water needs of different crops and whether those crops can tolerate recycled water.

With information on both water quality and end-use, it is possible to generate a minimum-water-requirement program for the area, sector, or subsector, broken down, if possible, by type or quality of water required. Requirements should be expressed as a range to make some allowance for periodic variation. Ideally, the program also includes information on temperature, salinity, or other water-quality considerations. A program for a hypothetical textile plant is shown in Table 1. The basic elements of this approach are outlined in Figure 1.

Comparing the costs of demand-side and supply-side options

The second major step in establishing a DSM program is to develop a conservation–supply curve, which describes the relative life-cycle costs of different

Table 1. Minimum water requirements for hypothetical industrial site A.

			Demand (m^3/d)		
				Potential	
End-use	Quality	Temp. (°C)	Current	Min.	Max.
Boiler makeup	Treated	≥20	250	200	350
Dye process I	Pure	75	120	75	120
Dye process II	Pure	95	135	125	135
Bleaching	Pure	50	140	100	140
Washing	Pure	65	240	120	350
Rewashing	85%	30	225	100	225
Ablution	Pure	20–35	125	100	125
Waste disposal	Reused	N/A	350	200	300

Note: Max., maximum; Min., minimum; N/A, not applicable; Temp., temperature.

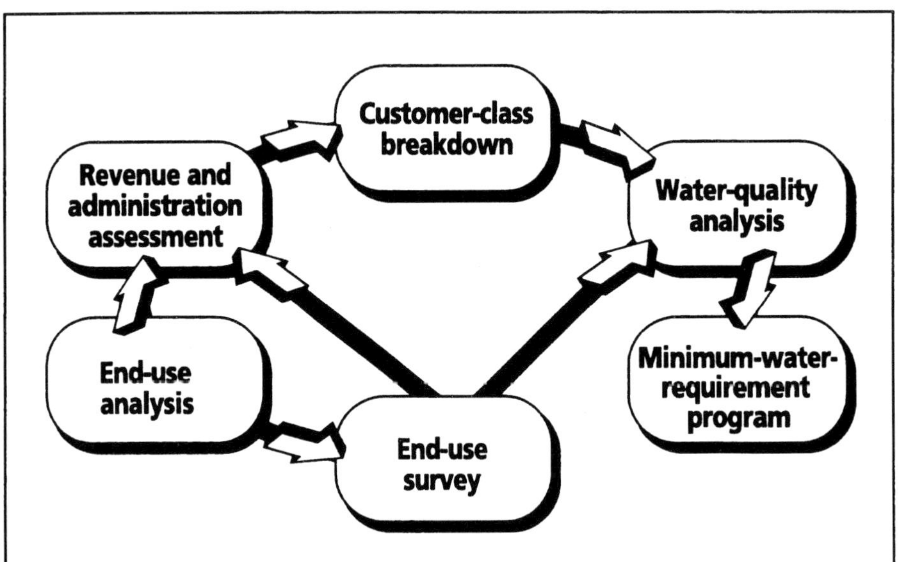

Fig. 1. Analysis of customer end-uses

conservation and supply options. For water, the comparison might be between, say, the cost per cubic metre of installing a drip-irrigation system and the cost of water delivered by a conventional spray-irrigation scheme or the cost of obtaining a new source of water.

In such an assessment, the supplier compares a range of conservation and supply options to see what the most advantageous investment would be, and at what point (if any) supply options are more cost-effective than conservation options. An exercise of this sort is fraught with many assumptions. To establish what the cost of a given conservation investment might be, for example, one must make assumptions about initial cost, product lifetime, operating costs, price elasticity of demand for the product and of the particular consumer group being targeted, and likely adoption rate of the product without incentives. Similar assumptions are made, of course, when one is planning a supply investment.

In the case of water, the choice of conservation technology is limited, as is the experience of consumer reaction to price or cost incentives. Technologies to consider include the following:

- low-pressure pipes, sprinkler systems, and drip systems for irrigation;
- water-recycling applications;
- improved lining materials to reduce seepage from canals;
- flow restrictors for domestic or industrial ablution;
- low-flush toilets;
- seasonal price variations; and
- leak-detection programs.

This analysis must reflect the true cost to local consumers of purchasing and operating a specific technology or application. It must also be updated periodically to reflect changes in local markets, tariff or tax effects, and cost of operation.

A difficulty in developing cost data is that many potential efficiency actions involve administrative or management improvements, rather than "technical fixes." In the case of energy suppliers, this problem is usually dealt with separately through information and training programs. These also represent a cost to the supplier and are intended to offset other supply costs. Perhaps the most important point about analysis of financial benefits of different conservation and supply options is that there must be an organization or agency responsible for providing water supplies, with discretionary authority over the course of future investments in the system and to which the benefits of conservation alternatives would accrue. There is no reason, of course, why this authority might not be a tribal or communal irrigation administrator, as opposed to a large riverbasin or dam authority or a municipal water-management authority. However, the development of DSM programs for water invariably implies an increase in the sophistication and technical competence of the management authority.

Disseminating information and training management

A major barrier to adoption of demand management is the provision of information dissemination and training. In a focused DSM program, therefore, water providers may be of assistance in two additional ways:

- disseminating information on efficient end-use technologies or better water-management strategies; and
- providing training on how to manage water resources more effectively.

DISSEMINATING INFORMATION — Dissemination of correct product information is essential if consumers are to achieve high end-use efficiencies. Many energy-DSM programs focus on information to the virtual exclusion of other kinds of subsidy or assistance and find that, if market conditions are correct (in particular, if the price of the resource is high enough), consumers will purchase efficiency independent of any further incentives. Such an assumption would obviously be inappropriate to the African situation, where consumers may be poor as well as ill informed. However, it is still essential to provide good information, not only on how to choose new technologies for water use, but also on how to manage the resource more efficiently by (in the case of agriculture, for example) applying better soil-conservation techniques, using water-recovery techniques, covering storage areas, or choosing less-water-intensive crops.

A pertinent example from the energy field in Africa is a project sponsored by the International Research Development Centre through the Energy for Development Research Centre in Cape Town. The project targets low-income populations in South Africa's urban and peri-urban areas. It aims to improve the flow of information about efficient and environmentally friendly domestic energy technologies (for lighting, cooking, space heating, and water heating) through a series of information workshops, efforts to improve appliance standards, and involvement of both the electricity utility and local authorities in the marketing and retailing of these technologies (EDRC 1993).

TRAINING MANAGEMENT — Training is the "flip side" of the requirement for better information. Examples from Egypt and Morocco show that improved on-farm management skills introduced through farmers' organizations can improve irrigation efficiency by 10–15% overall and productivity by as much as 30% (Xie et al. 1993). Management improvements can be fairly inexpensive if indigenous organizations and community groups are used as the delivery agents. From the viewpoint of water authorities and governments, this is a highly cost-effective investment in improving water-use efficiencies.

Choosing efficient technologies

In energy DSM, a distinction is often made between "prescriptive" and "audit-based" programs. The former involves provision of grants or subsidies to consumers who purchase certain types of energy-conserving equipment, such as energy-efficient motors or lights. The latter involves assessment of a single facility, such as an industrial plant or building, by a qualified consultant. The consultant then recommends a specific set of conservation investments, the costs of which can be shared by the utility and the facility owner. Prescriptive approaches are simpler to manage, although they involve careful market research and precise estimation of savings to the utility at different levels of adoption; audit approaches are more complex but much more precise in targeting potential savings.

Both programs are customer based rather than supplier based, and both involve selection of specific end-use technologies that have previously been studied by the utility and compared with other investments. Audit-based programs are generally preferred by utilities with limited capital because supply options can be deferred by adopting a small number of high-return conservation investments. However, audit-based programs require management sophistication and professional expertise, both of which are lacking in water authorities everywhere, not only in Africa. Prescriptive programs are preferred by utilities with major supply problems, substantial investment capital, and a good consumer-information database. In water management, prescriptive programs are appropriate where the water authority is trying to influence the behaviour of a large number of small customers, hence in the domestic sector and in small-scale irrigation. Such programs would prescribe specific end-use technologies, such as drip-irrigation systems for tobacco farmers or flow restrictors for domestic showers, and set a level of subsidy that will stimulate adoption by the user. In contrast, water audits directed to major industrial or agricultural users can almost certainly be justified financially. Water authorities can turn up significant efficiency opportunities through routine inspections of facilities.

Improving the efficiency of transmission–distribution

Transmission–distribution is a major area of potential improvement for electricity utilities, particularly where, as in much of sub-Saharan Africa, these systems are old and poorly maintained. Losses from these systems can range as high as 25 or even 30%, where 8–10% would be an acceptable figure in normal circumstances of short- and medium-range transmission. Apart from this substantial opportunity for efficiency gain, improvements in transmission–distribution efficiency have the additional benefit that they can be included in regular capital programs and can be managed entirely by the existing organization within the utility: they do not require special market-assessment programs or new-technology testing.

In the water case, the parallel is to replace or repair piping and canal systems in either municipal supply or irrigation. The evidence of efficiency losses in these applications is strong. Davies and Day (1986), for example, noted that known losses through leakage in the Cape Town municipal water-supply system (piping losses) were of the order of 14%. Xie et al. (1993) suggested that, for irrigation systems, there are extreme cases: unlined canals in pervious soils, for example, have transmission losses of 30–35%. With proper soil conditions, appropriate canal or pipe design, and better maintenance, it is possible to achieve transmission efficiencies of more than 95%.

The contrast between developing and industrial countries in terms of distribution losses is significant: whereas the level of distribution efficiency in developing countries overall is estimated at 68%, that of the United States is 78%, suggesting that a 15% improvement over the developing-country figure is theoretically possible. Accurate figures from Africa are not available, but it is estimated that distribution efficiency is even lower than the developing-country average, ranging from 40 to 60% (Xie et al. 1993).

The benefits from improvement in distribution are apparent, both for the end-user and for the provider of water. Actions such as relining canals, repairing and upgrading piping systems, replacing seals in pipeline pumps, minimizing leakage in storage tanks and reservoirs, and, if appropriate, reducing line pressures all require both capital investment and technical knowledge. They are best undertaken by central water-distribution authorities through direct investment, although it is possible that improved distribution efficiencies in irrigation systems could be encouraged by providing subsidies to users for lining local distribution canals or repairing pipes and gates.

Institutionalizing demand-side-management programs for water

Institutionalizing demand-side management of water requires that authorities responsible for water distribution have a clear set of responsibilities for maintaining secure and efficient water delivery, that they are responsive to price and quality issues, and that they have a reasonably detailed understanding of their customers' requirements. As this combination of qualities would be a rarity even in developed countries, it is more realistic to see the goal of institutionalizing DSM at water authorities as a process rather than an event. The status and capabilities of the customers themselves are key to this process. If water consumers cannot tolerate large variations in the price they pay for water (or any price at all) and if they have no opportunities to introduce efficient end-uses themselves, either financially or practically, then they are more likely to support an institutionally driven DSM program.

The final step in the institutionalization process is often the most difficult. DSM programs must be established by the very same organizations and individuals that are traditionally supportive of supply-side management. Existing water-management programs are characterized by a strong interest in developing new supplies and maintaining a given level of supply or quantity, and they are geared to dealing with expanding demand. To shift from this preoccupation to encouraging customers to reduce their demand for water and even to investing in these efforts is a major challenge that will almost certainly have to be supported by government itself, by nongovernmental agencies, and, possibly, by donors and lenders if it is to succeed.

Priorities for demand-side management of water resources

Establishing DSM programs for water resources requires that water-management authorities, as well as the donors and lenders who support water development, set entirely new planning priorities. Because for Africa the mix of organizations and authorities is quite different, the mix of priorities will be different from those of North America or Europe. They should include the following elements in rough order of priority:

1. *Improvement of the maintenance and operational efficiencies of existing delivery systems* — This is a major investment opportunity for water suppliers and must be encouraged. In large-scale systems, such investments should be driven by revenue yields: the returns on water sales must be sufficient to make it attractive for the supplier to invest in improving conveyance efficiency directly. In small-scale systems, such as traditional irrigation networks, indirect investment may be more appropriate because suppliers can subsidize the end-users' maintenance of their own part of the network and thus encourage a mutual concern for efficiency.

2. *Conjunctive management of groundwater and surface-water resources* — This is particularly important to avoid depletion of aquifers. However, the converse is also true: by using groundwater resources, it is possible to augment surface supplies and thus defer the cost of new surface reservoirs

and other capital-intensive supply systems. This is a strategy that can be realistically applied only at the level of the supplier or local or national government. Conjunctive-use strategies are particularly important in dealing with the crisis of urban water demand that afflicts many African cities, but they can also be applied in agricultural settings.

3. *Development of customer-oriented policies by large-scale suppliers, such as government itself or irrigation and riverbasin authorities* — This is a natural development for commercially operated utilities, but it will be more difficult for parastatal organizations, which tend to have a top-down orientation. Such a major shift in orientation can only be accomplished by significant changes in style of management. In other words, it will happen when management sees its services as customer driven.

4. *Maintenance of a realistic but generally cost-reflective pricing system* — This is the key to effective DSM, but it is not always possible to achieve it instantly or uniformly across customer classes. Nevertheless, most experts seem to agree that African agriculture, particularly commercial agriculture, is excessively cross-subsidized. They also believe that farmers will benefit in the long run (even though they may complain bitterly at first) by paying a larger share of the true cost of their supplies. Pricing must be tempered by the recognition that most African farmers are not involved in commercial agriculture. Similarly, most domestic consumers cannot afford substantial increases in water tariffs. Suppliers may, therefore, have to consider a "lifeline tariff" for these consumers and recover a larger percentage of revenue from those that can pay.

5. *Development of an end-use database* — This is critical for water suppliers, whatever their size and kind. Suppliers require detailed knowledge of how their consumers use water: what amounts and quality of water are required for specific end-uses. A user survey can accomplish this at relatively low cost, and the benefits of such an exercise far outweigh its costs. Above all, this kind of information will provide suppliers with a better indication of the end-use efficiency of their systems, as well as a more exact idea of how overall system efficiency can be improved by effecting technical improvements at the consumer level. In the African context, information of this sort is almost entirely lacking at present, except in a few relatively developed countries where metering systems have been used or where a high percentage of the population served falls within the formal economy.

6. *Provision of information on affordable water-management technologies* — This is a prerequisite for encouraging use of water-efficient technologies. African farmers, as well as domestic users, rarely have such access and, as a result, are not in a position to react rationally to price increases or even to direct conservation efforts. Dissemination of better product information has to come from either government or the supplier. It serves as an appropriate investment by them in increased efficiency of water delivery.

7. *Provision of appropriate financial incentives to consumers* — This is necessary if consumers are to increase efficiency: where possible, water suppliers should consider the benefits of investing directly in consumer efficiency. This is a key element in energy-DSM programs and will work just as effectively for water-DSM programs, even though the economics may differ because of water's universality and relatively lower transformation costs. Incentives can range from prescriptive measures, such as providing free flow restrictors to domestic consumers or establishing fixed subsidies for adoption of drip-feed irrigation, to audit-based measures that result in paydowns of efficiency investments for large industrial consumers.

8. *Increased water recirculation* — This is a customer-based activity that must be driven by pricing and information distribution. Suppliers need to identify areas where water from nonconsumptive uses can be diverted to consumers needing additional water at low cost. Just as in energy management obvious savings from recycling (often heated water) are overlooked by consumers because of low prices or lack of technical knowledge, so, too, in water management, recycling may need to be encouraged through either partial subsidies or increased information on recycling opportunities and demonstrations of its application in comparable situations. Recycling of sewage water is quite a different matter and demands strong central control and regulation.

9. *Improvement in the management of water resources* — This is particularly important for irrigation systems but may also be applied to municipal and industrial water management. Improved management skills can contribute significantly to efficiency gains and to overall system productivity in the case of agriculture, and it is logical, therefore, to expect authorities to invest in such improvements. Such investments could involve, for example, focused education programs for small farmers, management-assistance

activities delivered through agricultural extension programs, and consumer-information programs that stress the benefits of better management.

Recommendations for future research

It is clear that development of more-efficient water-management systems in Africa can be facilitated by application of some of the DSM principles employed in the field of energy. To achieve this, African governments, as well as the private sector, research organizations, and nongovernmental organizations, must act decisively to initiate specific reforms. In addition, they need to support continuing research in the field of DSM through existing university- and government-based research. The following areas are recommended for further research:

- development of appropriate end-use technologies for local conditions in Africa;
- analysis of the effects of pricing on consumer efficiency in different sectors;
- improvement in the accuracy and cost reduction of water-loss measurements;
- development of end-use databases for different customer groups;
- studies on the impact of transmission–distribution or conveyance losses on overall system efficiency; and
- development of conservation–supply curves for different authorities or sectoral applications.

References

Adams, W.M. 1992. Wasting the rain. Earthscan Publications, London, UK.

Barghouti, S.; Subramanian, A. 1991. Policy and management problems. *In* Barghouti, S.; Le Moigne, G., ed., Irrigation in sub-Saharan Africa. World Bank, Washington, DC. World Bank Technical Paper 123.

Brooks, D. 1994. Economics, ecology and equity: lessons from the energy crisis in managing water shared by Israelis and Palestinians. *In* Isaac, J.; Shuval, H., ed., Water and peace in the Middle East. Elsevier Science, Amsterdam, The Netherlands. pp. 441–450

Brooks, D.; Peters, R. 1988. Water: the potential for demand management in Canada. Report to the Science Council of Canada, Ottawa, ON.

Bryant, E. 1994. Water: tapping Africa's most basic resource. African Farmer, 12 (July), 25–28.

David, L.J. 1988. Environmentally sound management of the Zambezi riverbasin. International Journal of Water Resources Management, 4(2).

Davies, B.R.; Day, J.A. 1986. The biology and conservation of South Africa's vanishing waters. Centre for Extramural Studies, University of Cape Town, Cape Town, South Africa.

EDRC (Energy for Development Research Centre). 1993. Energy efficiency, equity and the environment: improving access to energy services for the urban poor of South Africa. Submission to International Development Research Centre. University of Cape Town, South Africa.

Elahl, M.; Khushalani, B. 1991. Technical issues of irrigation development in sub-Saharan Africa. *In* Barghouti, S.; Le Moigne, G., ed., Irrigation in sub-Saharan Africa. World Bank, Washington, DC. World Bank Technical Paper 123. pp. 69–82.

FAO (Food and Agriculture Organization). 1986. Irrigation in Africa south of the Sahara. FAO, Rome, Italy. FAO Investment Centre Technical Paper 5.

Frederick, K.A. 1993. Balancing water demands with supplies: the role of management in a world of increasing scarcity. World Bank, Washington, DC. World Bank Technical Paper 189.

Gellings, C.W.; McMenamin, J.S. 1993. DSM: what's real, what's not, and how we can tell the difference. Electric Power Research Institute, Palo Alto, CA.

Gleick, P.H. 1989. Climate change and international politics: problems facing developing countries. Ambio, 18(6).

Grainger, A. 1990. The threatening desert: controlling desertification. Earthscan Publications, London, UK.

Helming, S. 1993. Planning for the future: the cost of water. Paper presented at EDI Water Management Workshop, July 1993, Victoria Falls, Zimbabwe.

Kasimona, V.N.; Mkwaya, J.J. 1995. Present planning in Zambia for the future use of Zambezi River waters. *In* Matiza, T.; Crafter, S.; Dale, P., ed., Water resource use in the Zambezi Basin: proceedings of a workshop held at Kasane, Botswana. International Union for the Conservation of Nature and Natural Resources, Gland, Switzerland. pp. 35–43.

Lele, U.; Subramanian, A. 1991. Sectoral strategy and irrigation development in sub-Saharan Africa: some lessons from experience. *In* Barghouti, S.; Le Moigne, G., ed. Irrigation in sub-Saharan Africa. World Bank, Washington, DC. World Bank Technical Paper 123. pp. 60–68.

OECD (Organization for Economic Co-operation and Development). 1984. Rahad Irrigation Project performance report. OECD Project Performance Report 5130.

Schwarz, J. 1991. Sector water study in Israel. Presented at an international workshop on water resources management, June 1991, Washington, DC.

Scudder, T. 1989. Riverbasin projects in Africa: conservation vs development. Environment, 31(2).

Seckler, D. 1993. Designing water strategies for the 21st century. Winrock International Center for Economic Policy Studies, Arlington, VA. Discussion Paper 16.

South African Department of Water Affairs. 1986. Management of the water affairs of the Republic of South Africa. South African Department of Water Affairs, Pretoria, South Africa.

ter Vrugt, J. 1990. Private sector irrigation in Zimbabwe. *In* Barghouti, S.; Le Moigne, G., ed., Irrigation in sub-Saharan Africa. World Bank, Washington, DC. World Bank Technical Paper 123. pp. 88–90.

Wittfogel, K. 1957. Oriental despotism. Pergamon Press, New York, NY.

World Bank. 1990. The Gazira Irrigation Scheme in Sudan. World Bank, Washington, DC. World Bank Technical Paper 120.

——— 1992. World development report 1992: development and the environment. World Bank, Washington, DC.

World Resources Institute. 1992. World resources report, 1992–93. Oxford University Press, New York, NY.

Wyss, H. 1991. Prospects for irrigation development in sub-Saharan Africa. *In* Barghouti, S.; Le Moigne, G., ed., Irrigation in sub-Saharan Africa. World Bank, Washington, DC. World Bank Technical Paper 123. pp. 1–4.

Xie, M.; Kuffner, U.; Le Moigne, G. 1993. Using water efficiently. World Bank, Washington, DC. World Bank Technical Paper 205.

ALLOCATION OF WATER RESOURCES IN AFRICA: POTENTIAL FOR MOVING WATER IN AND OUT OF AGRICULTURE

Richard MacLean[1] and Joachim Voss[2]
[1]Agronomist and Consultant, IDRC; and [2]Anthropologist and Director, Sustainable Production Systems, IDRC, Ottawa, Ontario, Canada

Introduction

Africa has the highest growth rate in the developing world, and food production is not keeping pace with population growth (Winrock 1993). Some of the most limiting factors to improved food production on the continent are the quality and quantity of available water resources.

Because food security is so closely linked to water availability and irrigation in many regions, conflicts over water are evident. Increasing water scarcity will increase the potential for conflict within and between nations. Water security is further jeopardized by pollution from human and industrial activities. The fact that less than 1% of the Earth's aquatic endowment is freshwater that is reliably available for human consumption calls for greater management and conservation (Carruthers and Morrison 1994; FAO 1993a, b).

This report will focus on water-allocation problems in Africa. A comparative analysis of North Africa and sub-Saharan Africa will highlight water needs for the agricultural, municipal, and industrial sectors; water sources and irrigation schemes, as well as their potentials for increasing food productivity; the consequences of shifting water resources from the agricultural sector; and ways to increase water availability.

Irrigation

Irrigation has played an enormous role in increasing agricultural output worldwide. Because of increasing population, demand for water for the agricultural, industrial, and domestic sectors has grown significantly. Consequently, water allocation has undergone a significant shift since the turn of the century, when almost 90% of water resources were consumed by agriculture. Recent figures indicate that irrigation worldwide now accounts for 62%, whereas industrial and municipal consumption have increased from 6 to 25% and from 2 to 9% of allocation, respectively (FAO 1993a, b).

As the world population and competition for land increase, agricultural production on irrigated land will become more important. Today, an estimated 55% of global output of rice and wheat are from irrigated areas, and one third of the world's food is grown on 237 × 10^6 ha, or 17% of the irrigated arable land (Carruthers and Morrison 1994; FAO 1993a, b). Growth projections suggest that by 2025, 80% of global food production will come from irrigated land (FAO 1993a, b).

The population of sub-Saharan Africa is almost four times that of North Africa and is increasing at a faster pace (Table 1). The urban population is increasing much faster than the rural population.

Table 1. Demographic statistics for North and sub-Saharan Africa.

	North Africa	Sub-Saharan Africa
Population (million)	120	460
Growth (%)		
Rural	0.9	1.9
Urban	3.7	5.6

Source: FAO (1986a).

Sub-Saharan Africa lags far behind the rest of the world in proportion of irrigated arable land and its contribution to the total food supply. Of all the developing regions, water availability per capita is lowest in Africa. In North Africa, 11% of the land is irrigated (excluding Egypt's 100%), whereas in sub-Saharan Africa, it is only 3.5% of total cropped land (FAO 1986a, b, c). In North Africa, 3.4 × 10^6 ha is irrigated by large-scale systems, whereas in sub-Saharan Africa, half of the 5.3 × 10^6 ha is irrigated by large- and medium-scale systems; the other half, by small-scale systems. In terms of value, irrigation is responsible for an estimated 33 and 9% of the crops produced in North Africa and sub-Saharan Africa, respectively (Yudelman 1994).

Cereal production accounts for more than half the irrigated land in Africa, but in North Africa only 19% of the total cereal production is irrigated; in sub-Saharan Africa, a mere 2.5% (FAO 1986 a, b). The higher percentage in North Africa is due to Egypt's overwhelming contribution: cereal production covers 72% of Egypt's irrigated area. If the continent is to become self-sufficient in food production, land under irrigation will have to be increased.

Problems associated with irrigation include low groundwater potential, low aquifer-recharge rates, water-quality deterioration, and poorly drained, inefficient systems causing waterlogging and salinization. The latter two problems are the major causes of declining crop yields on irrigated lands in the arid and semi-arid regions (FAO 1986b). In many regions of North Africa, groundwater sources are being severely overexploited, and freshwater aquifers are deteriorating because of seawater intrusion (FAO 1993a, b). Thus, there is an immediate need to formulate and implement sustainable water-management strategies in this part of Africa.

Emerging scenario

Although the potential for expanding irrigation exists, agriculture will have to compete with the rapidly increasing urban sector, where municipal and industrial needs for water have dramatically escalated over the last two decades. In 1992, 88% of allocated water in Africa was for irrigation; 7%, for municipal use; and 5%, for industrial uses. Such a distribution among sectors is typical of low-income countries, where priority is given to agriculture because of the high multiplier effect a prospering agricultural sector can have on a developing economy. What is interesting to note, however, is that although high-income countries allocate a much smaller percentage of water to agriculture than do low-income countries, the total volume allocated to agriculture is still significantly greater because annual withdrawals per capita in richer countries are more than three times the volume consumed by low-income countries (Table 2).

Table 2. Sectoral water withdrawals by income group.

Country income group	Per capita annual withdrawal (m^3)	Per capita annual withdrawal (m^3)		
		Agriculture	Domestic	Industry
Low income	385	351 (91)	15 (4)	19 (5)
Middle income	454	313 (69)	59 (13)	82 (18)
High income	1 167	455 (39)	163 (14)	549 (47)

Source: World Bank (1992).
Note: Percentages in parentheses.

With urban populations continuing to grow, the allocation of water will have to change. The problem, however, is that if agricultural allocations are reduced and partially shifted to more lucrative urban markets, the rate of food production, which presently is not keeping up with population growth, may be further undermined. Can both objectives be met? Most believe they can. How they are met, however, is a controversial issue. The conservationists believe that by increasing water-use efficiencies of current irrigation systems, some of which are estimated to waste 60% or more of diverted or pumped water, enough water can be saved to meet urban needs without compromising food production. Although proponents of new developments believe that conservation is a critical issue, many feel that eventually the need for new sources will arise.

Management strategies

Our biological, environmental, cultural, and socioeconomic relationships with water distinguish it from most other natural resources. This unique status poses a formidable challenge to water-management authorities because policies must take into account legal, social, environmental, technological, economic, and political considerations (FAO 1993a, b). Water is less expensive than other resources, such as oil; consequently, policymakers worldwide have tended to emphasize supply management.

However, the idea of increasing the supply of water commensurably with rising demand from the agricultural, municipal, and industrial sectors has recently fallen into disfavour because of escalating construction costs, substantially lower cereal prices, a growing environmental and social awareness, and inefficient irrigation (FAO 1993a, b). Costs associated with new large-scale irrigation systems in Africa have dramatically increased and now range between 10 000 and 20 000 United States dollars (USD) per hectare, whereas medium-sized systems are estimated at 7 200 USD per hectare (FAO 1993a, b). With such exorbitant costs, even double-cropping of high-value products cannot be made profitable. The economic costs are high because of reliance on foreign experts, equipment, and supplies for construction. This, in turn, is associated with the lack or demise of an indigenous irrigation culture in much of sub-Saharan Africa. Furthermore, social analyses of the impact of large-scale schemes have revealed both official corruption and negative impacts on women and other less powerful social groups. Because of these socioeconomic complexities, water-management officials are now faced with reduced venture and public capital for water initiatives, curtailed irrigation subsidies, greater cost-recovery emphasis, and mandatory efficiency intensification (FAO 1993a, b).

The concept of increasing the price of water, as a policy instrument to alter water-consumption behaviour, is receiving greater attention. Using pricing policies in an effort to control water use may have far reaching implications that are intersectoral (agricultural versus industry), intrasectoral (rice versus maize), distributional (access and equity), and environmental (agricultural inputs and water quality) (FAO, 1993a, b). Determining where water is best used and its potential productivity is greatest is highly complex. Carruthers and Clark (1983) provided a classic illustration: although some 15 000 m^3 of water is required to irrigate 1 ha of rice, this same volume could just as well supply "100 nomads and 450 head of cattle for 3 years; or 100 rural families for 4 years; or 100 urban families for 2 years; or 100 luxury hotel guests for 55 days." The evaluation of these alternative uses remains somewhat subjective.

Although the price-control argument may have a sound economic basis, many arid and semi-arid countries remain reluctant to endorse it because water is so essential for life itself that it is not treated as a commodity. Cultural factors often play a pivotal role in water-allocation decisions, particularly where crop irrigation is essential to food security. In addition, increasing the cost of water may skew water distribution within society to the extent that only financially secure recipients could afford it. Such a scenario is often politically unacceptable because it is imperative that such a vital resource be equitably shared in society.

All of society relies on water for survival. As aggregate demand for water continues to soar, conflicts among user groups are bound to emerge. What are the consequences of shifting water allocation from agriculture to other sectors? Raising the cost of water may significantly increase the cost of irrigated produce, which in turn could affect a nation's competitiveness within a global economy. On a more regional scale, increased costs for water could be passed on to urban consumers via higher prices for irrigated goods. On the other hand, if growers are forced to absorb the increased cost of irrigated production, the tendency to shift to higher value crops is likely inevitable. What would the impact of such a shift in production be on, say, foreign currency earnings or the availability of traditional or local foodstuffs?

From the above statistics, it becomes obvious that although cereals account for a significant portion of irrigated land in Africa as a whole, only a very small fraction of total cereals in Africa is irrigated (19% in North Africa and 2.5% in sub-Saharan Africa). Consequently, there is an enormous potential to increase food productivity. The question is how? Strategies to tap this potential differ between North Africa and sub-Saharan Africa. In the latter, the irrigated area could be expanded by 330%. Although there is potential to expand irrigation in North Africa by an estimated 22% (Table 3), water sources are already overtaxed.

Table 3. Agricultural statistics for North and sub-Saharan Africa.

	North Africa	Sub-Saharan Africa
Food productivity, 1986 (%)		2.3
Cereal productivity, 1986 (%)		1.8
Irrigated cereals, 1986 (%)	19	2.5
Irrigated arable land, 1990 (× 10^6 ha)	7.6	5.0
Irrigated arable land, 1986 (%)	11	3.5
Potential increase in irrigated arable land, 2025 (× 10^6 ha)	1.6	16.5
Potential increase in irrigated arable land, 2025 (%)	22	330

Sources: FAO (1986a, b); IIMI (1994); and WRI, UNEP, and UNDP (1992).

North Africa

Irrigation systems in North Africa are generally inefficient and lack adequate drainage, resulting in severe waterlogging and salinization. Focus in North Africa should, therefore, centre on demand management, with conservation and increased water-use efficiency as the major policy objectives. The approach of repairing degraded systems and implementing stringent and effective conservation measures is gaining increased acceptance within the international community. This approach may be particularly appropriate in regions of North Africa that are under low and intermediate water stresses. Along with existing water supplies, estimated savings may then be reallocated to specific needs in other sectors. In the low- to medium-income countries, where 80% of available water is allocated to agriculture, it is estimated that a 10% increase in efficiency could translate into 50% more water for urban requirements (Yudelman 1994). Needless to say, such a strategy would also be less costly than new infrastructures, and society would benefit much sooner. The fact that present system efficiency can be improved at the local level is not disputed. However, some argue that at the watershed scale (the Nile River basin, for instance) little or no saving may accrue. Moore and Seckler (1993) commented that although increasing efficiency is commendable, additional water sources will inevitably have to be tapped.

Sub-Saharan Africa

In humid and sub-humid parts of the sub-continent, where climatic conditions are more favourable and water resources more abundant (FAO 1986a, c), the possibilities for increased irrigation are greater. In fact, 85% of sub-Sahara

Africa's irrigation potential remains untapped (FAO 1986a, c). The International Irrigation Management Institute estimates that irrigated land could be tripled. In those zones where rainfall is more evenly distributed, small-scale, community-based irrigation systems offer the greatest potential for significantly improving crop production. Supply-management strategies may be useful in sub-Saharan Africa, too, and improved water-use efficiency should not be ignored.

If a substantial increase in irrigated area is to be achieved on the continent, several limitations must be overcome. Traditional irrigation that has thrived in the past, notably in Zaire and Zimbabwe, should be revived. Large-scale irrigation systems are relatively recent. Before the middle of the 20th century, most water needs were met by capturing water from reliable and relatively inexpensive sources (FAO 1993a, b). A list of some of the advantages and disadvantages of large- and small-scale irrigation systems is provided in Table 4.

In sub-Saharan Africa, the most irrigated countries are Sudan (1.75×10^6 ha), Madagascar (960 000 ha), and Nigeria (850 000 ha) (FAO 1986b). Because most (65%) of the modern irrigation in the region is concentrated in Sudan, many of the remaining irrigation systems are traditional systems, operating much below their technical potential and efficiency. Improving the efficiency of traditional, small-scale systems calls for maximization of available rainfall and soil-management strategies that build water-holding capacities, promote greater water infiltration and percolation, reduce runoff, and decrease evaporation (mulching and conservation tillage).

Madagascar is one of the leading countries in sub-Saharan Africa in achieving irrigation potential (Yudelman 1994). Many of the systems in Madagascar have Asian farming components, and such systems may be relevant to, and beneficial for, other regions. If a small-scale traditional irrigation culture is to be reborn, Madagascar is the logical partner to share the knowledge it has acquired via community-based, farmer-to-farmer, technology-transfer mechanisms. Based on experiences from large-scale irrigation interventions in West Africa, such an approach may be more appropriate and yield better results.

Conclusion

In sub-Saharan Africa, the main challenge is to capture more of the available water resources. The main recommendations for the region are the following:

1. Look for ways to effectively build on local knowledge, institutions, and solutions for better water management. Revitalize indigenous systems and transfer experiences from Madagascar. All of these are promising avenues to be explored.

Table 4. Advantages and disadvantages of large- and small-scale irrigation systems.

Large-scale systems	Small-scale systems
Socioeconomic effects	
Regional economic growth	Increased local employment
Power generation	
Flood control	
Tourism	
Transport	
New = 10 000 – 20 000 USD/ha	New = 500 – 1500 USD/ha
Repair = 1 500 – 2 000 USD/ha	Little or no irrigation fees
Resettlement of inhabitants	Little or no resettlement
Increased cropping index	Protection from seasonal drought
Physical effects	
High level of water control	Low level of water control
High water-storage capacity	Low water-storage capacity
High water losses (≥60%)	Low water losses
Low efficiency	High efficiency
Environmental effects	
Fishing in reservoir	Less incidence of disease
Creation of wildlife sanctuaries	
Rerouting of watercourses	
Destruction of native habitat	Preservation of native habitat
Losses of forest and agricultural land in reservoir	Insignificant land losses
Increased waterlogging and salinization from high water table and no drainage	Increased waterlogging and salinization from high seepage in earthen canals
Degradation of riparian zone	
Increased flooding and downstream sedimentation	
Changes in water chemistry	
Eutrophication	
Barrier to fish migration	
Problems associated with changes in temperature	

2. Avoid large-scale irrigation schemes. The experience with such projects in the Sudan and in West Africa offers many lessons. There is now abundant literature describing the plethora of problems that have resulted from these donor-driven engineering solutions to water scarcity in the Sahel.

This is not to say that large-scale infrastructures may not be necessary or desirable under some circumstances, but rather that their necessity should not be assumed. When these infrastructures are deemed desirable by the majority of the potential users, particular care needs to be taken with the politics of implementation and management.

In North Africa, in contrast, most irrigation potential has already been tapped. Conflicts between urban and agricultural uses, as well as interstate conflicts concerning large river systems such as the Nile, are becoming severe. The main recommendations are the following:

1. Find alternative social and technical approaches to increase conservation and water-use efficiency in urban and agricultural settings.

2. Create effective and representative national and international institutions for negotiating water rights and resolving conflicts.

References

Carruthers, I.; Morrison, J. 1994. 2020 Vision: dramatic changes in the world agricultural and industrial production systems. *In* Feeding the world: issues and challenges for irrigated agriculture in the 21st century. International Irrigation Management Institute.

FAO (Food and Agriculture Organization). 1986a. Irrigation in Africa south of the Sahara. FAO, Investment Centre Technical Paper 5, Rome, Italy.

——— 1986b. Atlas of African agriculture African agriculture: the next 25 years. FAO, Rome, Italy.

——— 1986c. Annex IV: Irrigation and water control; African agriculture: the next 25 years. FAO, Rome, Italy.

——— 1993a. The state of the world: water policies and agriculture. FAO, Rome, Italy.

——— 1993b. Sustainable management of water resources for agricultural development. *In* World agriculture, pp. 110–112.

Moore, D.; Seckler, D., ed. 1993. Water scarcity in developing countries: reconciling development and environmental protection. Environmental Defence Fund and Winrock International Institute for Agricultural Development.

Winrock International. 1993. Agricultural transformation in Africa: round table discussion. Proceedings of the Seminar on Agricultural Transformation in Africa, Baltimore, MA. 1992, EPAT and Winrock International.

WRI (World Resources Institute); UNEP (United Nations Environment Program); UNDP (United Nations Development Programme). 1993. World Resources 1992–93.

Yudelman, M. 1994. Demand and supply of foodstuffs up to 2050 with special reference to irrigation. *In* Feeding the world: issues and challenges for irrigated agriculture in the 21st century. International Irrigation Management Institute.

WOMEN, MEN, AND WATER-RESOURCE MANAGEMENT IN AFRICA

Eva Rathgeber

Director, Regional Office for Eastern and Southern Africa, IDRC, Nairobi, Kenya

Introduction

By the early 1990s, nine African countries had per capita renewable water supplies of less than 1 000 m^3/year and were considered water scarce. The nine were Algeria, Botswana, Burundi, Egypt, Kenya, Libya, Mauritania, Rwanda, and Tunisia (Postel 1993). However, most other countries in Africa have large regions that suffer acute water shortages, either periodically or on a permanent basis. Consequently, in most of the continent, effective water-resource management is of critical importance.

In situations of scarcity, decisions about access to and use of water involve actors at the intergovernmental, governmental, regional, community, and household levels and often become highly politicized. The needs and perspectives of large- and small-scale farmers, of small- and medium-sized enterprises, of households, and of fisherfolk and others who earn their livelihood from water can differ significantly. At the same time, level of commitment of the different actors to conservation practices and to protection of water resources from contamination may also vary, and the question of whose interests prevail and receive top priority can create considerable tension.

At a global level, demand for water is increasing steadily, with a general trend toward diversification of use away from agricultural activities. Currently, about 70% of world freshwater resources are used for agricultural purposes, but with rapid global industrialization this is expected to decline to 62% by the year 2000 and to even less thereafter. Moreover, with increasing populations and improved living standards, domestic demand for water has grown significantly in all parts of the world, including Africa (Biswas 1993). In this context, the past decade has seen massive expansion of water projects in Africa. Although the African continent is still the least-irrigated and least-industrialized region of the world, sustained efforts continue to be made to provide safe and reliable water sources in rural and urban areas throughout the region, both for domestic consumption and for agricultural purposes.

This paper examines some of the concerns that have motivated African governments and donors to become involved with water projects. Although there is general recognition of the needs of "communities" for reliable water systems, it is argued that the different attitudes, perspectives, and needs of women and men with respect to water access and use have been given little focused attention by environmental planners and water-resource managers in Africa. More specifically, it is suggested that throughout the 1970s and 1980s, although concerted efforts were being made to increase water accessibility, little effort was made to integrate the economic roles of women into water-resource planning.

Drawers of Water (White et al. 1972), a classic American study of East African water issues, presented strong intellectual justification for donors' giving attention to the provision of domestic water to rural African communities and becoming aware of the sociopolitical issues related to conflicting needs and competition for water resources at the community level. The study became a standard reference work for water-resource planners in the 1970s and 1980s (M. Woodhouse, personal communication, 1995), but despite its specific focus on social issues, including water use, health, individual costs, and communities, it did not address the role of women in water-resource management, except to note that in most African societies it is considered women's work to carry water. White et al. (1972, p. 63) also noted that husbands "determine the arrangements, such as the site of the house or the investment of money in equipment, which make the job light or heavy." Their research implied that women were not significant decision-makers, even with respect to domestic water use and sourcing. In the ensuing years, this notion has been uncritically accepted, time and again, in the design of water projects around the continent.

Although research in the 1990s has demonstrated that African women are active participants in economic development, there has been relatively little systematic factoring of gender considerations into resource-allocation decisions. Despite substantial evidence of the economic profitability of this approach, traditional assumptions about the domestic roles of women continue to guide policymakers.

This paper suggests that the inherently conservative attitudes of environmental planners and water-resource managers toward mainstreaming gender concerns into their analyses reflect those of the broader society. It also suggests a continued, deeply rooted resistance to power sharing: this resistance finds its origins in a desire to maintain the status quo and to see women as passive bystanders. It provides a means of reducing political controversy around allocation of water resources because, by minimizing or discounting the specific needs of women, water-resource managers are able to remove at least one potential source of conflict over water allocation.

Water-resource planners, representing both national governments and donors, have focused their attention on women's reproductive, primarily domestic, responsibilities. This has been a fundamental error because women's strategic interest in water is closely associated with their productive roles (which, in turn, are often synchronous with their reproductive tasks). In overlooking this reality, planners have seriously undermined the capacity of women to make a substantial contribution to national-development processes, and this has led to significant losses of water resources in Africa.

The paper begins with an examination of the attitudes of donors and governments toward water-resource management. A number of inconsistencies and points of conflict between these attitudes and user needs are identified and discussed. The paper then examines the extent to which women have input into water-resource management, focusing on their needs with respect to both their reproductive and productive roles.

Inconsistencies and points of conflict

Although donors and governments often use the term *gender* when addressing different needs with respect to water resources, they usually mean *women*. Male needs and priorities provide a standard against which female (or gender) interests are measured and often minimized. A true gender focus on water-resource management would entail a complete rethinking of the issues of access to water, of the differing and sometimes conflicting interests of men and women, of their relative input into and power over decision-making, and of the role and composition of the community. These are difficult and potentially contentious questions, and it is not surprising that they receive little attention from water engineers, who assume that the provision of water facilities is the primary task at hand and that related social issues will sort themselves out over time.

As noted, governments and donors usually see women's involvement in water-resource management primarily from the perspective of their roles in social reproduction. Green and Baden (1995) cite numerous examples from World Bank documents about women's role in "providing, managing and safeguarding water" for use by the family. Because household water provision is still a female responsibility in most African societies, especially in the rural areas, donors and governments have tended to assume that women's strategic interest in water is concentrated primarily in having access to convenient, reliable, and safe sources close to the homestead. Further, because women are seen as having first-line responsibility for the maintenance of family health, they are thought to have a special interest in, and responsibility for, water and sanitation. Available evidence suggests that, although both assumptions have some validity, they fail to capture women's equally pressing needs for water to enable them to engage in economic production, whether in agriculture, in microenterprise, or in other income-generating areas.

Government planners rarely factor microanalysis of daily water needs into national water-resource plans. To some extent this is a function of the operational models of most African governments. There are often weak linkages between ministries of economic development and ministries looking after the welfare of women and children. Ministries of economic development (often responsible for water-resource-allocation planning) tend to focus primarily on the attainment of overall water objectives and less on the needs of, or impact on, different sectors of the population (Yoon 1991).

Both governments and donors recognize that water-resource planning must be multisectoral, but water projects are usually designed to meet technical rather than social objectives. Although it is expected that installed water systems will be used by local populations, engineers rarely examine the pattern of needs of different groups of potential users and take this information into account when designing the system. It is usually assumed that social groups will change their habits of interaction to accommodate and take advantage of available water, rather than its being recognized that newly available water will not be used optimally because this does not conform to the existing norms of social groups. Because of this lack of sensitivity to prevailing cultural patterns, it should not be surprising that water projects often fail, especially after donor support is withdrawn.

The Lesotho Highlands Water Project provides a large-scale example. The project includes four large dams, water-transfer works, and a hydroelectric power plant. Funded by the World Bank, the European Economic Community, the European Investment Bank, the United Nations Development Programme (UNDP), several bilateral agencies, and the Government of South Africa, the project's objective is to divert water from the Orange and Senqu rivers in Lesotho to service South African industry (Jayseela 1993). The project will destroy between 1 and 1.3% of Lesotho's arable land, which is significant because of the overall shortage of arable land, and substantial grazing land will also be appropriated. It would appear that the needs of South African industry have prevailed over those of local farmers, but because Lesotho has a very high proportion of female-headed households, it is likely that women will be disproportionately affected.

Because women's productive roles usually are not taken into consideration by macrolevel environmental planners, it is not surprising that women were apparently overlooked by the planners of the Lesotho project. But even at a household level, prevailing power relations between men and women usually ensure that the water needs of women receive lower priority than those of men, although the economic contribution of women's work can be of significant importance to the household. As discussed below, women's access to water resources is often strictly controlled, both directly by their husbands and indirectly by existing cultural norms and practices.

Most interventions in the water sector have focused primarily on expanding the supply or availability of water for given populations or on serving specific needs, like irrigation or industrialization. Water-resource planners have assumed that demand exists and that it will adjust equitably to new sources of supply. Disproportionate attention has been focused on issues of delivery and on short-term indicators of achievement (for instance, finishing on schedule and within budget and providing for certain quantities of water flow). Commonly, it has been assumed that populations will be equitably served and that individuals and groups will have access as required.

This has proven to be mistaken, and there is considerable anthropological evidence to suggest that often populations tend neither to be equitably served nor to adjust their water-related practices to take full advantage of new sources of water. To illustrate, it is useful to examine women's water needs in the context of their reproductive and productive roles. It should be stressed, however, that the bifurcation of women's responsibilities into reproductive and productive ones is artificial — women's work tends to encompass both types of responsibilities within a single context. For example, household food processing, which is done primarily for family consumption, often creates a surplus that is sold outside the home and becomes an important source of additional family income. Because of this, the reproductive and productive roles of women routinely become merged.

Women's reproductive roles

The term *reproductive* is used here in the sense of social rather than biological reproduction. It refers to all of the services provided by women to ensure the healthy maintenance of their families, including cooking, cleaning, and child care. Because reliable and convenient access to potable water is important in helping women fulfill these tasks, donors and governments often assume that women's primary strategic interest in water relates to their domestic roles. Significant research on women and water has been undertaken from this perspective, focusing especially on

- access to water;
- decision-making related to water use; and
- family health and water and sanitation.

Access to water

An average household in developing countries consumes about 40–60 L of water daily for drinking, cooking, cleaning, personal hygiene, etc. Meeting this need usually entails several trips for women and children to water-collection points, sometimes involving several hours. In some mountainous regions of East Africa, women spend up to 27% of their caloric intake in fetching water (Lewis 1994). Many traditional rural water sources have become contaminated as a result of human and animal waste and agricultural runoff. Especially during the dry season, rural households often collect their water from contaminated sources.

In urban areas, the situation is not necessarily better, and when household cash resources are meagre, it may be worse. For example, in Nairobi, slum dwellers buy water from vendors or collect it from communal water points, which are often highly unsanitary, and in Mombasa more than 60% of slum dwellers have no access at all to clean water (Government of Kenya and UNICEF 1992). In Kumasi, Ghana (as in many other African cities), water connections are shared by numerous families, and hygienic conditions are inadequate. The poorest families are forced to purchase water from more prosperous neighbours who have access to water connections (Whittington et al. 1993). Generally, in slum areas in African cities, there is considerable water contamination. Human and other waste is dumped into streams and drainage ditches or dries and becomes airborne, thereby creating a serious health hazard.

Although there have been important exceptions, small-scale water-supply projects aimed primarily at satisfying domestic needs have not been a priority among bilateral donors (Dankelman and Davidson 1988). There are many examples of failed small-scale water-supply projects in Africa. These projects often have been developed in response to specific political agendas, and, consequently, selection criteria for their location have been neither rigorous nor need driven. Projects have commonly been started with inadequate information bases concerning the needs, preferences, and level of commitment of the community (Harnmeijer 1993). Because political capital is at stake, there has often been no real attempt to gauge the interest and commitment of potential beneficiaries. Consequently, when donors have withdrawn, sometimes there has been little interest at the community level in maintaining the system. Moreover, the costs of continuing maintenance and repair have sometimes not been accurately judged by donors, which means that community members may be unable or unwilling to make the necessary financial contributions to keep the new system smoothly functional. These are critical considerations in the sustainability of water projects because specific expectations exist on the sides of both the donors and the potential beneficiaries. For example, a study in Tanzania found that local water systems were not maintained by the communities because they considered this the continuing responsibility of the agency that had installed the system. Villagers had provided labour to install the pipes, but they considered maintenance as being the duty of the agency (Waters 1992). This underscores the fact that, before water-supply projects are implemented, there should be in-depth interaction with local communities, including women, to ensure that expectations and responsibilities on both sides are clearly stated and understood. As discussed below, it is not uncommon for donors to assume that men speak on behalf of the community, because women are hesitant to voice opinions in public meetings or fora.

Research on the use of installed water systems has shown that women consider various factors when deciding where they will obtain water. Water planners often take a somewhat linear approach to water-sourcing, thinking that women (and men) will make choices based exclusively on cost–benefit analyses in terms of time, convenience, and water quality. However, numerous other factors can come into play, and at different times of the year and for different intended purposes, water choice may be driven by a set of unquantifiable perceptions. In some cases, for example, time spent queuing for water at a village pump may be perceived as a welcome opportunity to share gossip and information and to socialize; at other times, use of the village pump may be rejected because of pressure of other obligations that mitigate against the necessary time commitment. Studies in Egypt and Ghana showed that women did not use available public water points or pumps because of the waiting time involved. In both cases, time and convenience influenced decision-making more than water quality did (El-Katsha and White 1989; Kendie 1994).

Sustainability of water and sanitation systems is often problematic in the absence of year-round use of the systems. In some cases, to save time, women will use closer sources of water, even if the water quality is not optimal. In other cases, the financial contributions needed to maintain the system may become too burdensome for communities after donor support has ended, and systems may remain in disrepair for long periods (Narayan-Parker 1988). Lack of local expertise to repair and maintain systems also continues to be a problem, as does unavailability of spare parts.

Decision-making related to water use

Cost recovery and willingness to pay have become considerations for donors in the 1990s, particularly for the World Bank in lending to the water sector. Results with cost recovery in rural areas have been mixed.

Community water projects have appointed committees to collect money from households using the installed water systems, but cash contributions for water-system maintenance have been especially difficult to collect over the long term. It tends to be more problematic to collect money if alternative traditional sources of water are close by or if the water produced by the system is of poor quality and quantity (Narayan-Parker 1988). Gender analysis can lead to some insights because decisions about how household cash resources will be spent are often made by men (who may not have a strong commitment to contributing to water-system maintenance because collection of water is not their responsibility).

In urban areas, householders often pay for a water connection or buy water from informal-sector vendors. Again, in cases where women do not have their own income or control over how it is used, decisions about the purchase of water are commonly made by male heads of households.

In recent years, the World Bank has undertaken considerable research on how much families are willing to pay for individual access to potable water, especially in urban areas. A 1990 study in Morocco found that poor urban households placed a high priority on access to potable water, ranking it in importance immediately after food and clothing as a necessity for which they would be willing to pay (McPhail 1993). A Nigerian study showed that in Onitsha, householders regularly paid 3–5% of their income for water (Whittington et al. 1991). During the dry season, the poorest households paid as much as 18% of their incomes to purchase from private water vendors, and the researchers concluded that on an annual basis, households in Onitsha pay private vendors more than twice what it would cost to operate and maintain a piped water system.

Because most of the World Bank studies did not examine differences in the views of men and women, it is difficult to determine whether gender-based preferences were present. Studies that have desegregated gender attitudes toward payment for water have reported mixed results. Green and Baden (1995) cite evidence from Tanzania and Haiti that suggests that, in general, women were more willing than men to pay for access to public taps, whereas in India and Nigeria, women were not prepared to pay as much. Women's limited decision-making influence over household finances may make them reluctant to suggest greater expenditure on improved access to water, particularly if this expenditure would primarily benefit the women by easing the burden of their household work.

Cultural patterns may also have an impact. Research in Tanzania revealed that even when water was readily available, women hesitated to use it to keep their children's faces clean because this was not considered a priority and went against the wishes of their husbands (McCauley et al. 1992). A study in Kenya (Government of Kenya and UNICEF 1992, p. 91) noted that "there is evidence that in some communities men will give precedence to the building of a corrugated iron roofed house, purchase of a bicycle or marrying of a second wife over the supply of basic house hold necessities such as food and water." It is evident that prevailing cultural role expectations for both men and women can have critical importance in determining attitudes toward water-resource use and management.

African women's capacity to have input into water-resource use and management is further hampered by their lack of exposure to science and technology. To make informed choices and decisions, potential beneficiaries of water systems should have basic knowledge and understanding of the technologies involved. Women are particularly disadvantaged because of their lack of confidence about technological matters and because of negative male attitudes toward female technical knowledge.

Donors have made many attempts to address this issue, particularly in the context of women's reproductive responsibilities. For example, a workshop on water and sanitation, held in Kenya in 1987, emphasized the need for women to be involved in the choice and transfer of water-related technologies. It was stressed that women should have input into decision-making related to the use of individual or collective water technologies; they should be educated on the scope of available technologies so that they can make appropriate choices and recommendations; and they should be consulted about the cultural implications of different technologies (INSTRAW 1987a). It was strongly recommended that women be given appropriate training to allow them to participate in water- and sanitation-related decision-making at all levels. United Nations agencies and bilateral donors sponsored many similar workshops in the mid- to late 1980s in an attempt to erase prejudices about the involvement of women in the repair and maintenance of water and sanitation systems.

In Botswana, the Swedish International Development Agency assisted with the establishment of water points in rural areas throughout the 1980s, and by the early 1990s more than 80% of the country's rural population had reasonable access to safe water. However, an evaluation study showed that household-use patterns had changed relatively little. Water-related hygiene practices in rural households were still poor, and households still fetched water, on average, seven times daily. In retrospect, it was recognized that the project had failed to take into account the traditional roles assigned to men and women in Botswana (Simpson-Herbert 1992, 1994). First, women were not involved in the original planning of the Botswana water projects. The projects were seen as community-based ones and it was assumed that women's input would come through the community, but in reality the cultural practice was for men to make decisions on behalf of the community. Second, once the water-supply systems were in place, women became involved as operators, rather than as more highly paid managers or professionals (these positions were reserved for men). Third, few women were given the opportunity to seek professional training. They were kept in the clerical pools. The evaluation found that prevailing cultural roles made it highly unlikely that women would seek technical training and that the women themselves had low confidence in their ability to perform well in technical jobs. The overall effect of these combined factors was to marginalize female participation in the water projects, despite the fact that women were considered as having strong interests at stake in the establishment of rural water systems. Women were expected to contribute labour, but they had little real power or input into decision-making.

In contrast, the South Coast Handpump Project, in Kenya, developed by PROWWESS and UNDP in collaboration with the Kenyan Ministry of Water Development, emphasized female participation in pump repair and maintenance from the beginning. This undoubtedly was a critical factor in the project's success (Narayan-Parker 1988). Initially, the project had been intended exclusively to train women for pump repair and maintenance, but the female participants themselves requested male participation, arguing that young women would get married and move to other communities. Interestingly, both men and women preferred to have women serve as treasurers for water committees — both sexes believed that women were more trustworthy with money. The success of this project underscores the importance of flexibility and responsiveness to local cultural patterns as an element of development planning and implementation.

In general, however, although projects and workshops aimed at teaching technical skills to women have addressed important deficiencies and problems, they have also tended to reinforce rather than challenge the notion that women should have primary responsibility for ensuring the availability of water for domestic purposes. As such, they have not questioned the gendered power relations and role expectations of women and men in communities or promoted the idea of joint responsibility for water and sanitation. In this sense, they have not gone far enough in changing attitudes.

Family health and water and sanitation

Health education has been an important entry point for women's participation in water projects. Although measurement of the effects of good water and sanitation on health is methodologically complex, there is some evidence that when there is significant improvement in a community's level of water and sanitation, the impact on health can be substantial (Government of Kenya and UNICEF 1992).

As a result of the sexual division of labour, women are often at higher risk of exposure to waterborne diseases. Washing clothes, bathing children, drawing water from surface sources, and, in some regions, working in flooded rice fields all increase rural women's risk of exposure to disease-ridden water sources (Lewis 1994). Cultural preferences, shared by men and women, can also have a negative impact. A study in Egypt showed that women (and men) believed that canal water lathered clothes more effectively and made them whiter, so they preferred to wash their clothes in the canal (where risk of exposure to schistosomiasis was high), rather than washing them under a tap or in a washing machine (El-Katsha and White 1989).

African governments and donors have tended to see the maintenance of family health through safe water and sanitation as primarily a female responsibility. For example, in 1987 the United Nations International Research and Training Institute for the Advancement of Women held seminars in Ethiopia and Kenya that focused specifically on women, water supply, and sanitation. These seminars moved beyond a purely functional perspective on women's involvement because they emphasized the need for women to become key actors in water supply and sanitation, with input into decision-making and knowledge about technical maintenance of local water and sanitation systems. At the same time, however, they focused heavily on women's role in cleaning the environment and in looking after family hygiene and on educating women about water and sanitation issues (INSTRAW 1987a, b). To some extent, this reinforces the notion that women have a special responsibility for family sanitation and hygiene. Such efforts do not address the more fundamental issue — that water and sanitation facilities, which are used by both men and women, should be a joint responsibility, not only in terms of sharing of labour input but also in terms of sharing of decision-making and associated power.

Some African research has suggested that, even when the potentially negative health effects of poor water and sanitation systems are known and understood, neither women nor men necessarily give them a lot of consideration when making decisions about water use. For example, a Tanzanian study showed that although women understood the relationship between clean faces and trachoma, they did not wash their children's faces more regularly when water became easily accessible (McCauley et al. 1992). In Kumasi, Ghana, public latrines were often not used because they were inconvenient (Whittington et al. 1993). A study in Tanzania found that villagers did not consider it their duty to maintain water pipes provided by a donor agency, and they were completely unswayed by the clean-water–good-health argument because it was inconsistent with their own world-view (Waters 1992). This would seem to suggest that public-health messages about the relationship between clean water and good health not only need to be reconceptualized but also must be addressed to both men and women. In other words, maintenance of family health should not be seen as solely a female responsibility.

African women's groups have often been used by donors to implement water and sanitation projects. Frequently, income-generating activities have been added to the projects to make them more attractive to women. However, experience has shown that when such projects become successful they are sometimes called "community projects," and men take credit for them (Ong'wen 1993). In this way the community label can be potentially damaging for women. Similarly, views that are supposedly held by the community may in fact reflect the opinions of a few of the wealthier, most influential individuals (usually men).

Discussion

The perspectives of donors and governments and of women and men can be quite divergent on issues of control and decision-making in domestic water access and use. First, it seems clear that, at the village level, decisions are not necessarily made according to either a linear or a Western, "rational" model. Although it might seem self-evident to a donor representative that maintenance of good health should be a top priority, this may not be so in the mind of a woman burdened with the need to complete a multitude of daily tasks. Her priority might be to save time or to save money; therefore, she might choose to use contaminated water for which she does not have to line up or to pay.

Second, although donors may assume that cultural practices will be changed to take advantage of water that has been provided, this is not true. Cost–benefit analyses (both in terms of economic cost and in terms of time and convenience) do not capture the symbolic importance that people may attach to prevailing norms and practices, and this symbolic importance may well be the decisive factor when choices are made.

Third, the motivations of men and women to pay for potable water may differ, and in an unequal power relationship, it is more likely that the interests of men will prevail. Because provision of water is considered a basic duty of women, the lightening of that task will not necessarily be given a high priority by men, who make the decisions about expenditure of household incomes.

Finally, it would seem that donors have increasingly tended to assign responsibility for water-system maintenance and repair to women, under the assumption that women have the more strategic interest in ensuring operational efficiency to enable them to meet their domestic needs. Consequently, donors have developed "women's" water projects and sponsored workshops, seminars, and short courses aimed at breaking down stereotypes about women and technology. This has introduced women to aspects of water technology and exposed them to some basic technical skills. However, it has also reinforced the idea that women should continue to take primary responsibility for meeting household water needs and maintaining family health, especially in relation to waterborne disease. In this sense, it reinforces the existing status quo and the notion of socially reproductive labour as being almost entirely a female responsibility.

Women's productive roles

Rural African women need access to water resources to carry out their reproductive responsibilities, but they also need access to water to undertake work with direct economic benefits for themselves and their families. Sometimes their needs are in direct conflict with those of male members of their households (although at other times interests will be shared or complementary).

Interestingly, there is very little literature about women's need for equitable access to water to enhance their capacity to earn livelihoods and contribute to household incomes. The importance of water as a key input into economic production, especially in agriculture, is obvious and has been extensively discussed in the literature on water-resource management in Africa. But the question of access to water is seldom desegregated by gender. This is most apparent when one examines economic production outside the agricultural sector.

Women's participation in irrigation schemes

Irrigation is not as widely practiced in Africa as in the Middle East and other regions. However, existing projects have tended to focus primarily on the development of appropriate irrigation hardware. Progress is measured by the amount of land put under irrigation, rather by than the extent to which farmers make effective use of irrigated land and the consistency of the irrigation system in place with prevailing social practices and patterns (Diemer and Vincent 1992). Not surprisingly, the success rate for large-scale irrigation projects in Africa has been low (Brown and Nooter 1992).

For the most part, women have been omitted from both large- and small-scale irrigation schemes in Africa, often because of land-tenure issues. Zwarteveen (1994) suggested that irrigation settlement schemes in Africa and elsewhere have usually been based on three assumptions:

- Male heads of households control farm resources and labour.
- Improved incomes for male farmers will lead to improved quality of life for the entire household.
- Farm households are composed of nuclear families (man, woman, and children).

It is clear that these assumptions have had inadequate empirical verification.

The establishment of irrigation systems in Africa has been marked by some of the same inadequacies already noted in the design of large-scale water projects. Baseline information about the prevailing gender division of labour in agriculture, farming practices, and land-tenure systems has rarely been collected and carefully analyzed before irrigation systems are established. The three above-mentioned assumptions about family structures and household resource allocation have often guided irrigation planners. One consequence has been the dispossession of female farmers. Research on the Blue Nile Irrigation Scheme, established by the British in Sudan in 1954, found that women traditionally had the right to own land in the region of Wadal Abbas. When the irrigation scheme was established, land was taken away from existing farmers, including women, and reallocated exclusively to men. There is strong evidence of a decline in female farming as the irrigation schemes became widespread (Bernal 1988). Similarly, when irrigated rice farming was introduced in the Jahaly Pacbarr Project, in Gambia, resource and access rights of women declined. Although women benefited from the increased economic prosperity of the area, they became more dependent on senior male heads of households, providing labour for their land, whereas in the past the women had usufructuary rights of their own (Carney 1988). An irrigated rice project in Cameroon was unable to pay for itself because women, who were not assigned land but were expected to work in their husbands' fields, withheld their labour in order to grow sorghum for family subsistence outside the irrigation scheme (World Resources Institute et al. 1994). In Kenya, the Mwea Irrigation Scheme appropriated all available land, investing its control in the hands of the scheme managers, who were men. Women lost rights to land they had traditionally used to grow food crops for subsistence. Consequently, women were forced to turn to their husbands for cash to buy food and became more dependent on men than they had ever been in the past (Zwarteveen 1994).

A recent study of smallholder irrigation schemes in Zimbabwe found that irrigating households had higher yields of cash crops but that few female-headed households belonged to irrigation schemes (Chabayanzara 1994). A primary reason for the low representation of female farmers was the prevailing government policy of allocating irrigation plots exclusively to men. In fact, in the Zimbabwean study, those women who headed irrigating households tended to be de jure heads, mostly widows. De facto female heads of households were more prevalent in dryland farm areas because the lower cash incomes earned from such farming practices encouraged men to migrate to urban areas to seek alternative employment. It would appear that traditional, conservative views about the role of women, held in this case by government officials, made it impossible for female farmers to gain entry into the scheme, except by default. Therefore, female farmers had no choice but to continue to practice dryland farming.

Research in Kenya on smallholder rice irrigation in the Kano Plains revealed similar inequities (Hulsebosch 1993). Most women were not active members of the water users' associations in the rice schemes, and those who did attend meetings were not allowed to speak before men or to express opinions in opposition to those expressed by men, despite the fact that women performed up to 61% of the requisite labour in their own and their husbands' plots. Even when both men and women participated in irrigation schemes, their needs and priorities sometimes differed. Women were less interested in night irrigation because cultural norms made it difficult for them to work after dark. Men were interested in having watering places for cattle; women in having communal areas for washing clothes and dishes. These different perspectives were not effectively represented by the water users' associations because women were underrepresented and were not given an equal voice in decision-making. The study also found that the water guards in the rice-irrigation area were exclusively male, and men generally tended to receive more water from the irrigation schemes.

Because there is considerable evidence to suggest that women are not equal partners and beneficiaries in smallholder irrigation schemes, it is not surprising to find that women may sabotage irrigation efforts or withhold labour as a form of protest. In Kenya, Cameroon, Burkina Faso, and Gambia, there is evidence that women have withheld their labour from or reduced their labour input to their husbands' irrigated plots if they felt that they were receiving insufficient compensation for their work (Zwarteveen 1994). Similarly, Brown and Nooter (1992) pointed out that women often withdraw from irrigation schemes because the cash benefits from participation go to their husbands (who are the official holders of the land and the official participants in the schemes).

Women's work outside the agricultural sector

Aside from agriculture, the informal sector is Africa's biggest single source of employment. In all sub-Saharan countries, both rural and urban women are involved in petty trading, selling of cooked food, and brewing of ale and beer or other traditional drinks. Other informal-sector activities typically undertaken by women include running tea kiosks; processing and selling street foods, like rice balls, roast maize, or groundnuts; producing handicrafts; selling charcoal or firewood; tailoring; and making dresses. In some cases, women prefer to participate in group enterprise because it offers protection against interference or manipulation by husbands or male relatives (Jiggins 1989). However, a Ugandan study found that rural, female small-scale entrepreneurs tended to be isolated and to work independently (Kyomuhendo 1992).

Commonly, the activities undertaken by women are extensions of their domestic roles, and women often operate directly from their homes, sometimes relying on assistance from their children. Most of these businesses require a low initial capital outlay, but access to water is often essential for both production and sanitation. There appears to have been little analysis of the importance of access to water in women's choice of particular informal-sector business activities, in the success or failure of their businesses, or in the capacity to expand their business activities. A study of urban agriculture in Nairobi revealed that urban food production is an important source of family food and additional income for women but that women's access to irrigation was minimal (Freeman 1993).

A study of women's petty-commodity production in rural Uganda revealed that economic necessity, either that of providing basic support for their families or that of supplementing inadequate incomes of their husbands, was the basic motivating factor for participation in informal-sector economic activities (Kyomuhendo 1992). In no case was economic independence or a general desire to improve socioeconomic status a primary motivating factor. More work needs to be done to verify these findings in other countries, but it would seem logical that women engage in informal-sector work primarily because they need the additional income to sustain themselves and their families. If this is indeed the case, then there is a strong argument to be made for incorporating their needs for access to water for economic production into water-resource planning and for assigning such needs the same high priority as assigned to the needs of male small-scale entrepreneurs.

The importance (or lack or importance) of access to water resources for women's informal-sector activities raises a set of interesting research questions, which fall outside the usual concerns of both donors and governments.

Discussion

Water-resource planners, governments, and donors all seem to share the view that water associated with direct economic production (agriculture or industry) has a higher value than water used for domestic purposes because there is a possibility of greater income accruing to the users as a result of the former. This approach reinforces the tendency to negate or devalue women's work and economic contributions — it assumes implicitly that the time women spend in fetching water for domestic use is not productive time. In other words, the social reproduction activities of women are assigned no economic value, despite the fact that if women did not perform such services, families would have to pay substantially to purchase them in the open market. Of course, this is equally true in other regions of the world, but in the African context it takes on a particular urgency because of the tendency for reproductive and productive work to merge both inside and outside the household.

Development economists and planners have resisted any suggestion that women's unpaid labour for family survival should be factored into national accounting systems, and there seems to be an unspoken agreement that such work is a natural, necessary, and unquantifiable part of the human cycle of reproduction. Economic valuation of women's reproductive work would somehow imply a denigration of "core" human values. It is beyond the scope of this paper to fully explore this issue; however, it is relevant to point out that most of the tasks performed by men are, in contrast, assigned an economic value. It is unlikely that African economists will begin to factor women's reproductive labour into national accounting systems in the near future, but there is certainly a strong argument to be made, at the microlevel, for sharing reproductive tasks between women and men and, at the macrolevel, for economists' giving women's economic aspirations and contributions a priority equal to that of men's.

Areas for further research

A number of areas for further research have emerged. It is clear that there is need for further empirical studies about actual household patterns of decision-making related to water use. To what extent do women's attitudes and perceptions differ from those of men? Are there significant gender differences in attitudes toward paying for water? Do power relations between men and women within the household make it less likely that women will argue strongly for use of cash reserves to pay for water?

A second key area for research concerns water-project decision-making at the macrolevel. To what extent do planners recognize that the perspectives and needs of women and men may differ and that the needs of women go beyond their reproductive tasks and have an economic value? To what extent are microneeds taken into account? There is a need to develop effective methodologies that capture microneeds of various categories of users (including women) in such a way that this information can be effectively used by water-resource planners.

A third important set of issues relates to the cultural biases in gender-role expectations and attitudes toward water access and management. To what extent do tradition and culture circumscribe the capacity of women to make effective use of available water resources for both reproductive and productive needs? What possibilities exist for breaking down some of the existing cultural barriers?

A fourth area for further research relates to an examination of policy and legal instruments that discriminate against women's full access to water resources. How can laws be changed to ensure that women are not excluded from irrigation projects? What are the water needs of informal-sector entrepreneurs?

Conclusion

The analysis presented suggests strongly that development work in the area of water-resource management in Africa has tended to build on traditional views about the roles of women. Although there has been increased emphasis on ensuring that women receive training in water-pump repair and maintenance and in organizing women's groups to manage village water systems, priorities have been set with the assumption that women's strategic interests lie primarily in the fulfillment of their reproductive roles. Donors have played an important part in reinforcing such stereotypes about women's needs, interests, and priorities. There has been continuing emphasis on the water needs of the family and on the maintenance of adequate sanitation to ensure family health.

Few analysts have argued that it makes good economic sense, from an efficiency perspective, to ensure that female farmers and small-scale entrepreneurs have the same access to water as male farmers. By minimizing the importance of women in economic production, water-resource planners have effectively removed a potential source of conflict. Competition over water resources in Africa is already intense, and by viewing the water needs of male and female farmers as essentially homogeneous and by accepting the role of men as spokespeople for the entire community, donors and government planners have reduced the number of actors who have a stake in decision-making related to water-resource management. It is not suggested here that this is being done intentionally but that it is a predictable outcome of the kinds of traditional assumptions about the role of women still held by donors and government planners. Is this significant? Would it actually make a difference if gender were routinely factored into water projects?

The literature suggests that in some cases women have sabotaged irrigation schemes because they felt that their interests were not adequately represented. They felt neglected or disenfranchised when land over which they formerly had control was appropriated for irrigation projects. Inevitably, it is men who have the greatest voice on irrigation-water committees and women who are expected to provide labour to ensure the success of their husbands' farms. It would seem that women do employ resistance strategies if they feel that their interests are being overlooked, but clearly this is an area that requires further empirical research.

Again, it would appear to be self-evident that in some cases, water will play a more critical role in the success or failure of a small business than in others. The most significant point to be emphasized here is that women's work must be taken seriously by water-resource managers. Women's work, both inside and outside the household, is essential to family maintenance.

Integrating gender issues into water projects in a systematic way will involve a rethinking of social structures and organizations. Planners will have to go beyond a vaguely defined concept of community and commission baseline studies that reveal actual decision-making patterns and access to resources within social groups. It is the tension between the needs and priorities of these different actors that forms the basis of gender analysis. In essence, this will involve a rethinking of existing social structures and organizations. But to do otherwise would be to perpetuate approaches that fail to take into account the needs of all members of the community. It is not enough to integrate women's concerns. They must become a central part of the water agenda (Yoon 1991), with the same legitimacy and urgency as the concerns of any other major group of users. It is necessary to recognize, of course, that women as a group are far from being homogeneous. Like men, they are divided by social class, by ethnic group, and by differential access to resources. Consequently, it would be a major error to assume that there will be solidarity in women's views and, indeed, experiences concerning water-resource management.

References

Bernal, V. 1988. Losing ground: women and agriculture in Sudan's irrigated schemes: lessons from a Blue Nile village. *In* Davison, J. ed., Agriculture, women and land. Westview, Boulder, CO.

Biswas, A.K. 1993. Water for agricultural development: opportunities and constraints. Water Resources and Development, 9(1), 3–12.

Brown, E.P.; Nooter, R. 1992. Successful small-scale irrigation in the Sahel. World Bank, Washington, DC. World Bank Technical Paper 171.

Carney, J.A. 1988. Struggles over land and crops in an irrigated rice system in the Gambia. *In* Davison, J. ed., Agriculture, women and land. Westview, Boulder, CO.

Chabayanzara, E. 1994. Small holder irrigation development: impact on productivity, food production, income and employment. *In* Steven, A.B., ed., Issues in African rural development 2. Winrock International, Arlington, VA.

Dankelman, I.; Davidson, J. 1988. Women and environment in the Third World: alliance for the future. Earthscan Publications, London, UK.

Diemer, G.; Vincent, L. 1992. Irrigation in Africa: the failure of collective memory and collective understanding. Development Policy Review, 10(2), 131–154.

El-Katsha, S.; White, A.U. 1989. Women, water and sanitation: household behavioral patterns in two Egyptian villages. Water International, 14, 103–111.

Freeman, D.B. 1993. Survival strategy or business training ground? The significance of urban agriculture for the advancement of women in African cities. African Studies Review, 36(3), 1–22.

Government of Kenya; UNICEF (United Nation's International Children's Emergency Fund). 1992. Children and women in Kenya: a situation analysis. Government of Kenya and UNICEF, Nairobi, Kenya.

Green, C.; Baden, S. 1995. Integrated water resources management: a gender perspective. IDS Bulletin, 26(1), 92–100.

Harnmeijer, J. 1993. Selecting villages for impact in water supplies: an expanded role for the baseline study. Health Policy and Planning, 8(3), 224–231.

Hulsebosch, J. 1993. Evaluation of measures to enhance a gender balance in water users' associations: experiences in the Southwest Kano Project in the Kano Plains in Kenya. Report to the Small-holder Irrigation and Drainage Project, Ministry of Agriculture, Nairobi, Kenya.

INSTRAW (United Nations International Research and Training Institute for the Advancement of Women). 1987a. Women, water supply and sanitation. A national training seminar, 9–13 Nov. 1987, Nairobi, Kenya.

——— 1987b. Women, water supply and sanitation. A national training seminar, 23–28 Nov. 1987, Addis Ababa, Ethiopia.

Jayseela, F. 1993. An ethical diversion? The Lesotho Highlands Water Project. Droplets, Newsletter of the African Water Network, 3.

Jiggins, J. 1989. How poor women earn income in sub-Saharan Africa and what works against them. World Development, 17(7), 953–963.

Kendie, S.B. 1994. Survey of water use behaviour in rural north Ghana. *In* Jacqueline, S. ed., Women, health and the environment: an anthology. WHO, Geneva, Switzerland.

Kyomuhendo, G.B. 1992. The role of women in petty commodity production and commerce: a case study of rural peasant women in Uganda. Eastern Africa Social Science Research Review, 8(2), 1–19.

Lewis, N. 1994. Safe womanhood: a discussion paper. International Federation of Institutes for Advanced Study, Toronto, ON.

McCauley, A.P.; West, S.; Lynch, M. 1992. Household decisions among the Gogo people of Tanzania: determining the roles of men, women and the community in implementing a trachoma prevention program. Social Science and Medicine, 34(7), 817–824.

McPhail, A.A. 1993. The "five-percent rule" for improved water service: can households afford more? World Development, 21(6), 963–973.

Narayan-Parker, D. 1988. People, pumps and agencies. The South Coast Handpump Project. PROWWESS/UNDP, New York, NY.

Ong'wen, O. 1993. Women and community participation. Droplets, Newsletter of the African Water Network, 3.

Postel, S. 1993. Water scarcity spreading. *In* Brown, L.; Kane, H; Ayres, E., ed., Vital signs: the trends that are shaping the future. The World Watch Institute, New York, NY.

Simpson-Herbert, M. 1992. Women and water. World Health (July–Aug.).

——— 1994. Gender and management issues in the water sector: the case of rural Botswana. African Water Network Newsletter, 1.

Waters, T. 1992. Lifeworld and system: of water systems and grain mill development in rural Tanzania. African Studies Review, 35(2), 35–54.

White, G.F.; Bradley, D.J.; White, A.U. 1972. Drawers of water: domestic water use in East Africa. University of Chicago, Chicago, IL.

Whittington, D.; Donald, T.L.; Kyeongae, C.; Jeffrey, A.H.; Venkateswarlu, S. 1993. Household sanitation in Kumasi, Ghana: a description of current practices, attitudes, and perceptions. World Development, 21(5), 733–748.

Whittington, D.; Lauria, D.T; Mu, X. 1991. A study of water vending and willingness to pay for water in Onitsha, Nigeria. World Development, 19(2/3), 179–198.

World Resources Institute; UNEP (United Nations Environment Program); UNDP (United Nations Development Programme). 1994. World resources 1994–95. Oxford University Press, New York, NY.

Yoon, S. 1991. Water for life. Paper presented at the Symposium on the Impact of Environmental Degradation and Poverty on Women and Children, 27–30 May 1991, Geneva, Switzerland.

Zwarteveen, M.Z. 1994. Gender issues, water issues: a gender perspective to irrigation management. International Irrigation Management Institute, Colombo, Sri Lanka, Working Paper 32.

PART II

Subregional Contributions

BETWEEN THE GREAT RIVERS: WATER IN THE HEART OF THE MIDDLE EAST

David Brooks
Director, Environmental Policy Program, Environment and Natural Resources Division, IDRC, Ottawa, Ontario, Canada

Introduction

Water has been the key natural-resource issue during the three millennia of recorded history in the Middle East. Some regions of the world are drier, and others have higher populations or larger economies, but no other region of the world embraces such a large area, with so many people striving so hard for economic growth on the basis of so little water.

Three dimensions, three crises

This paper describes water stress in the region between the Nile and the Tigris–Euphrates river systems and extending southward to encompass the Arabian Peninsula and the Gulf Islands, a little bigger than what is sometimes called the Mashrek. Reference will be made to a larger group of countries that includes the Maghreb, Libya, Sudan, and Turkey. Throughout this region, the origin of water stress is not limited to scarcity but stems from three interacting crises:

- Demand for fresh water in the region exceeds the naturally occurring, renewable supply.
- Much of the region's limited water is being polluted from growing volumes of human, industrial, and agricultural wastes.
- The same water is desired simultaneously by different sectors in some society or wherever it flows across (or under) an international border.

Water scarcity has been a source of stress since history began, but water quality is a new problem coming to dominate the crisis in many parts of the world. In this region, though, the politics of water is probably of greater concern than anywhere else in the world. Moreover, because these three crises are interdependent, any resolution must deal with all three — quantity, quality, equity — at the same time if it is to be economically efficient, ecologically sustainable, and politically acceptable.

Physical and economic sources of stress

For most countries in the Middle East, water is the limiting resource for development (Fig. 1). Iran, Iraq, Lebanon, Sudan, Syria, and Turkey are all fairly well endowed with water; the three Maghreb countries (Morocco, Algeria, and Tunisia), Israel, and Egypt form a middle group; and Jordan, Libya, and countries of the Arabian Peninsula are least well endowed. For Palestinians, the West Bank is relatively well endowed with water resources (Lowi 1993; Lonergan and Brooks 1994). (Much of the water that occurs in the West Bank is today used in Israel.) The Gaza Strip is perennially short of water. However, water availability per capita is decreasing in every country of the region (Fig. 1).

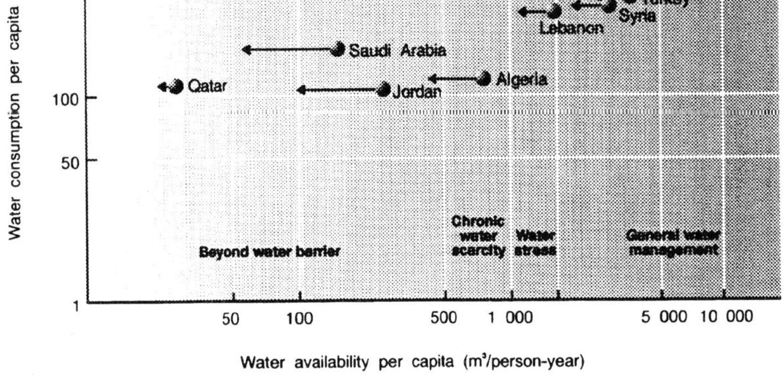

Fig. 1. Water-stress codes for selected Middle Eastern countries (circles depict water stress in 1988; arrows show the changes in per capita water availability by 2020, assuming that per capita consumption stays the same). Source: Falkenmark et al. (1989).

Variation and uncertainty

The region between the Nile and the Tigris–Euphrates is highly varied in geography and climate. Coastal plains merge in a few kilometres with mountain ranges, which then plummet to rift valleys with the lowest land elevations on Earth. Rainfall ranges from more than 1 000 mm/year to essentially nil. The average is about 250 mm, which is the limit for unirrigated agriculture, but in this region, averages can be highly misleading. It is much more important to understand the spatial, seasonal, and annual variations in rainfall than national or annual averages.

Bakour and Kolars (1994) showed that the Mashrek lies in a transition zone. To the north, the land receives more rainfall; to the south, even less. The dominant hydrological characteristic is the combination of aridity and uncertainty. Figure 2, taken from their study, compares the two curves: one showing diminishing average rainfall, and the other showing increasing variance (both from north to south across the region). In their words, "the zone of greatest unpredictability is at the intersection of the precipitation and variance curves," that is, in the populated, semi-arid regions of the Middle East. Whereas regions of higher rainfall sometimes suffer droughts and regions of lower rainfall sometimes experience floods, this region has to cope with both.

Even where the variations of rainfall are predictable, sharp transitions bedevil any attempt to use averages. Rainfall along the coast, at the higher elevations of many countries, and in the northern part of the region is more than 500 mm/year, which suggests that irrigation is unnecessary. However, all the rain falls in four winter months, so storage systems are necessary to hold back the flow and permit release during the summer, when demand for water is at its peak. (Unfortunately, large storage systems are generally built as surface reservoirs, which increase evaporation and, thus, decrease still further the volume of water available.)

Variations also occur across short distances. The north end of the Gaza Strip gets rain at upwards of 400 mm/year; barely 50 km to the south, at the border with Egypt, the Gaza Strip gets less than 250 mm/year. The thin coastal strip of Lebanon gets nearly 2.5×10^9 m^3/year; just 50 km to the east, across the Lebanon mountains, the Beka'a Valley, where most of the irrigated agriculture is located, gets only 0.9×10^9 m^3.

The most important variations in rainfall are neither seasonal nor geographic but annual. In eastern North America, reliable flow (defined as what can be expected 9 years out of 10) will be 60–80% of the long-term average; in western North America, reliable flow falls to 30% of the average. In the Middle

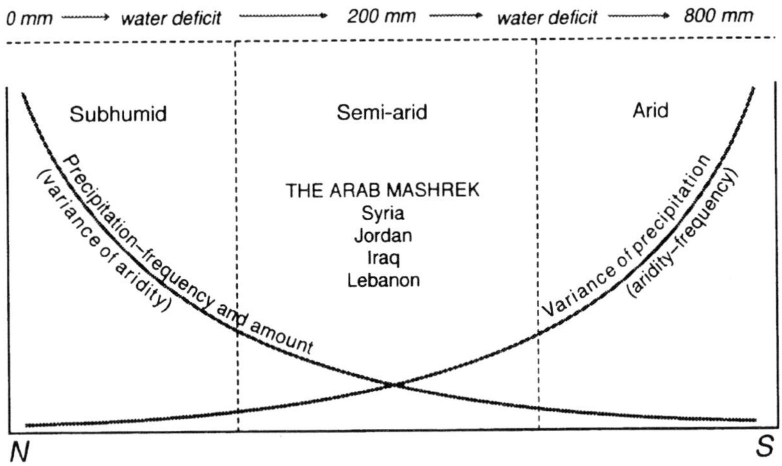

GENERAL CHARACTERISTICS

enough water → usually enough land		← enough land never enough water
seasonality (warm/cold) drought	fluctuations–variability runs of good and bad years extreme seasonality	extreme aridity floods
low yields per unit area →		← low yields per person
dispersed populations →		← concentrated population
control of land → (traditionally by cavalry)	raiding	← control of water (fortified towns)

TRADITIONAL DEVELOPMENT *(key words: kismet, tradition)*

rainfall agriculture	dry farming/nomadism	hydraulic works
transhumance (conversent up and down mtns)	village/tent symbiosis	qanats/wind towers, etc.

MODERN DEVELOPMENT *(key words: productivity, balance of payments)*

mechanize	settle the nomads/get rid of goats	make the desert bloom
tractors, new seed, monoculture	enclosure, policing, reforestation	deep drilling/ big dams wholesale irrigation desalination, air conditioning

Fig. 2. General characteristics of development in the Arab Mashrek. Source: Bakour and Kolars (1994).

East, reliable flow is less than 10%. These year-to-year variations in rainfall in the Middle East have enormous implications for water systems. In contrast to Europe, Canada, and much of the United States, extreme years in the Middle East must be treated as normal, not abnormal, and water planning and management must focus on risk minimization, not maximum use.

Demography and economy

Most countries in the region are experiencing rapid population growth, with rates of 2.5% per year. Although population densities are not particularly high by world standards, density per hectare of agricultural land is another story. Bahrain has a remarkable 7 000 people per arable hectare; Egypt and Kuwait have close to 2 000; and the other Gulf States, Israel, Jordan, and Lebanon, have more than 500 (Rogers 1994). In the United States, the ratio is less than 2. Other sources of stress come from rapid urbanization (which increases the demand for high-quality water, without diminishing the demand for irrigation water), and the booming economic growth.

Demographic and economic sources of stress are common to many regions of the world. What makes a difference in this region is the dominance of agricultural uses of water, mainly irrigation. Even in relatively urbanized Lebanon, irrigation takes close to 80% of total water. Every country provides water to farmers at moderately to heavily subsidized prices. Although not all water withdrawn for irrigation is actually consumed, the proportion that is returned to a watercourse and the extent of degradation in the natural recycling remain controversial (Moore and Seckle 1993). In Egypt, a great deal is returned with relatively little degradation, but it is dangerous to generalize from the Nile (Allen 1994).

In contrast to agriculture's dominance of regional-water accounts is its decline in economic accounts. Agriculture represents less than 5% of gross national product (GNP) in Israel and Turkey and less than 10% in Jordan and Lebanon. Agriculture represents about 20% of GNP in Egypt and Iraq, and a little more in Syria and the West Bank. Only in Ethiopia, the Gaza Strip, and Sudan does the share approach 40% of GNP. In most countries, the share of employment in agriculture is higher than agriculture's share of GNP, but in Israel and Lebanon, with their capital-intensive farming, the share of employment is lower. With such disproportionate use of water in one sector, and a declining sector at that, political sources of stress are bound to occur.

Quantity: the economic crisis

With its limited water resources, the region from the Nile to the Euphrates not surprisingly contains some of the most parsimonious users of water in the world. A Bedouin may get along with as little as 4 or 5 L/person per day for all uses. Only Australian Aborigines seem to use less. What is a surprise is that the same region also contains some of the least parsimonious users of water. The Gulf States rate among the highest per capita users of water in the world.

Almost all the states of the Arabian Peninsula are consuming much more water than their annual renewable water supply, as are Israel, Jordan, and Libya. Egypt, Syria, and Sudan are fast approaching this situation. Indeed, some projections suggest that by 2025 domestic uses (about 100 L/person per day), plus municipal and industrial uses, will require all the freshwater available, leaving none for agriculture, in the countries of the Lower Jordan (Shuval 1992; Assaf et al. 1993). Even if no more water is devoted to agriculture over the next few years, these countries are in trouble: their water use is unsustainable, which implies that their whole economy is unsustainable. Apart from desalination or imports, the only ways to significantly improve the situation are to improve water efficiency in existing uses and to shift water from low-productivity to high-productivity water sectors. The dominance of irrigation means that both efficiency improvements and sectoral shifts must emphasize agriculture.

Principal sources of water

Throughout history, this region has depended on three main sources of water: rivers, aquifers, and imports (through trade in food). Allen (1994) estimated that the quantity of water imported indirectly into the Middle East as food amounted to 50×10^9 m^3, equivalent to one third of the water directly used and about equal to the annual flow of the Nile in Egypt. Of course, the Middle East also exports food, so the net indirect trade in water is smaller.

Rivers are the best-known sources of water, and the rivers in this region include two of the greatest in the world, the Nile and the Euphrates. As well, many short streams or ephemeral wadis occur, typically being fed by springs in the mountains and spilling into the sea. Aquifers of various types are also common. Some are replenished regularly by rainfall and constitute renewable resources; others contain water buried in sediments eons ago and, thus, constitute nonrenewable resources; a few others occur along fracture zones. Over time, as surface sources have become fully committed and as technology has permitted deeper drilling, there has been a shift to groundwater. Even so, only about 10% of the total supply for the region comes from groundwater. However, in Israel and Jordan, the share from groundwater approaches 50%; and in the Arabian

Peninsula, 100% (if desalination is put to one side). Apart from a few cities, no other region is so dependent on aquifers.

Few opportunities remain for further development of major rivers in these countries; on the contrary, if development occurs, it will more likely be in the upstream countries, such as Ethiopia and Turkey, which could reduce flows downstream (Allen 1994; Hillel 1994). Major freshwater aquifers remain to be developed (indeed, to be discovered), but they are either very deep or located far from points of consumption.

Recycling water

Recycled water may well be the fourth conventional source. Today, the use of treated, recycled sewage water is accepted practice in countries, such as Egypt, Israel, Jordan, and Morocco. (In some countries, raw sewage continues to be used, which risks the spread of cholera and other diseases.) Countries in the region that are truly short of water will have shifted largely from fresh to recycled water for irrigating crops early next century. Some recycled water receives only primary treatment, in which case, use should be restricted to nonfood crops, and special care should be taken to protect farm workers. A good part of the water receives secondary treatment, which means that it can be used for crops that are eaten after cooking. Only water that has received tertiary treatment can be used for all crops. The European Economic Community is considering a proposal to embargo crops grown in reclaimed sewage, but this rule should be resisted as a nontariff barrier designed to protect European farmers. If implemented, it would be a severe blow to Middle Eastern agriculture. Tests for heavy metals and other contaminants not eliminated by conventional treatment are reasonable, but not a flat embargo.

Alternative sources of water

There are many other small, but potentially much larger, sources of water. They can be divided into two groups, depending on a pair of criteria that tend to move in parallel: capital requirements and degree of centralization. Of particular interest in the Middle East are the following:

Low-capital–decentralized solutions

- rainwater catchment from roofs and other structures
- rainwater harvesting in fields and in limans
- capture of flood and winter runoff
- desert dams
- aquifer recharge

High-capital–centralized solutions

- desalination of seawater
- desalination of brackish water
- imports of water by tanker, pipeline, or medusa bags
- cloud seeding

Much more attention should be paid to the low-capital–decentralized options than to the high-capital–centralized ones. To a large extent, the former are not only technically proven but typically more cost effective, given the marginal costs of new conventional water supply. Some options, such as rooftop rainwater catchment, produce only small total quantities of water, but it is potable water. Even in areas of low rainfall, such as the Gaza Strip, it is possible to design low-cost systems, with cisterns scaled to families, that will provide for all drinking and cooking needs (5–7 L/person per day) in most years. The greater problem is not designing the systems but convincing people unused to this technique that the stored water is, indeed, potable.

With the exception of cloud seeding, which has been practiced in some countries for years, the remaining systems are generally too expensive for widespread use. A partial exception must be made for brackish-water desalination, which, depending on location and salt content, can be an appropriate option. Imports of water must be considered, if only because in some countries the use of water depends on the importation of energy. For example, about 20% of Jordan's electricity and 12% of Israel's is used for water pumping. Although such a high share may not be typical (both nations must move large volumes of water from lower to higher elevations), these countries are willing to import oil, in one case, and coal, in the other, so that they can pump water. Of the import options available, the Canadian technology of medusa bags (large plastic bags towed behind ocean-going tugs) appears attractive but remains to be proven at full scale. Turkey is the one country in the region that appears to be both willing and able to consider water imports.

Desalinated seawater is, of course, the ultimate, unlimited source, and about two thirds of the world's "desal" capacity is located in this region. However, all technologies use such vast quantities of energy that major use is restricted to those countries with low-cost oil reserves or heavy-oil fractions left after refining. More potential exists for desalination of brackish water containing up to about 5 000 ppm of salts. Relatively small quantities of brackish water have been desalinated in many countries. In its peace treaty with Jordan, Israel committed itself to desalinating the saline springs that it has diverted to the Lower

Jordan, which now makes this water too salty for use by Palestinian and Jordanian farmers.

A few sources do not fit neatly into either of the two groups. For example, there are numerous aquifers that contain 1 000 – 5 000 ppm of salts — too salty to be potable but acceptable for certain uses. Some of these aquifers are huge, including one that underlies almost the entire Sinai and Negev deserts. Issar (1994) suggested that the area could be developed on the basis of industrial and agricultural uses of saline water. The aquifers contain fossilized water and, therefore, are nonrenewable, but the supplies are so vast that centuries of use is possible at any likely pumping rate.

The submarine springs that occur all along the coast of Lebanon, Israel, and Gaza and maybe farther are another source of unknown potential. The locations are well known to fisherfolk because some fish can be found at the point where the freshwater and seawater mingle, but no one has ever proposed a practical and ecologically safe method for capturing and bringing the water to the shore.

Main uses of water

Except in a few cases where the objective is to preserve natural beauty, water is not desired for its own sake but because it can satisfy human needs. Despite this basic fact, nowhere is the information available on water use as detailed or comprehensive as that on water supply. Worse yet, with some exceptions, the information is based on deliveries of water, not actual use, which means that it is impossible to make accurate measures of efficiency.

Water use can be broadly divided into four categories: household, municipal, industry, and agriculture. (Municipal use refers both to the water delivered to commercial buildings and hotels, mainly for the same purposes as households, and to the generally larger amounts used for municipal gardens, street cleaning, fire fighting, etc.) Household use accounts for 3–20% of the consumption; municipal, 3–10%; industry, 1–10%; and agriculture, 50–90%.

A surprisingly small proportion is required to meet the drinking-water standards defined by the World Health Organization. At 5–7 L/person per day, only about 2×10^6 m^3/year is needed for every 1 million inhabitants, which is not very much. In most parts of the world and, certainly, in this region, major uses do not require water of potable quality. Depending on the potential contact with humans or the possible fouling of equipment, water of moderately to significantly lower quality can be used. In the future, cities may move to dual systems, with a small pipe providing potable water for drinking and cooking and a larger one providing lower quality water for other uses. Unfortunately, in many countries of the region, there is only a single system, and the water delivered meets only the lower standard.

Pricing reform is at the top of the agenda of every economist who has looked at water supply and demand in the region. Studies of water use in the region have found that consumers of water are subsidized and that these subsidies increase water use above what it would be if consumers had to pay the full costs. Despite the differences in supply, most consumers in this arid region pay no more for water than consumers in humid parts of the United States (Rogers 1994). The need for water-pricing reform is reinforced by other factors. First, because the volumes of water needed to sustain life are so small, it does not matter much whether drinking water is subsidized in the interests of equity or public health. Drinking water is not the problem! Second, among all consumers, farmers receive the greatest subsidies, particularly in comparison with value of output. For some crops, the value added by irrigation is less than the average cost of supplying the water. Third, water prices are typically compared with average cost, but, for economic efficiency, consumers should be paying the marginal cost (the cost to get additional water), which is higher yet. Although most other countries also subsidize water consumers, especially farmers, few countries are so short of water as those between the Nile and the Euphrates. Gradual movement toward more economically efficient water-pricing should be possible without major social losses but with definite ecological gains.

Conservation of water

Conservation of water, including both increases in efficiency of existing uses and changing use patterns, has always been a major consideration in the water-short Middle East. (*Efficiency* and *conservation* are commonly used as synonyms, but, more precisely, the former refers to minimizing inputs to achieve a given output, whereas the latter includes changes in the output. Less formally, efficiency deals with how one accomplishes a task, whereas conservation also includes changes in the task.) However, as we learned from deeper analysis of energy use, the fact that a region is short of energy (water) does not imply either that existing uses are efficiently satisfied or that the pattern of use is appropriate. Many factors, including capital barriers, ill-designed policies, inaccessible technology, lack of information, and habits and traditions intervene.

Continuing with the analogy with energy (Stiles, this volume), we can say that private firms and public bodies must begin to look at reductions in the use of water as a source of supply — a very large source, equivalent to but, in many ways, better than new primary sources. In separate analyses of water-rich Canada and water-short Israel, quite comparable proportionate cost-effective savings were identified (Brooks and Peters 1988; Kahana 1991). This does not mean that Canada and Israel are at equivalent levels of water efficiency. Canada uses four and a half times as much water per capita as Israel. However, because water prices are so much lower in Canada than in Israel, the potential gains in economic efficiency are similar. Without changing use patterns but relying only on off-the-shelf technologies, savings typically exceed 25% and, in some cases, reach 50%. In just 2 years, the city of Jerusalem cut its municipal water use by 14%. These results suggest that the largest potential source of water for most countries in the region will be found through savings achieved by conservation.

Although none of the countries under study comes close to maximizing economic, much less technical, potentials of efficiency in using water, the dominance of irrigation requires special attention. Israel is generally regarded as a model of efficiency in irrigation. Israel pioneered the development of drip irrigation and has gone on to improve the technique with sensors and computer controls that respond to plant requirements, rather than using a predetermined watering schedule. Today, water use per irrigated hectare is 40% less than it was in 1955, and gains continue to be made, although at a declining rate (Kahana 1991; Hillel 1994). Drip irrigation also reduces the likelihood of both salinization and pollution from runoff. Other nations in the region, notably Jordan, have adapted drip irrigation for many crops and are now producing their own pipes and other equipment. Unfortunately, drip irrigation is a capital-intensive technology, and it is not appropriate for all crops.

The greater question about Israeli agriculture and, by extension, all agriculture in the region is not, however, whether water is used efficiently in irrigation but whether irrigation is an efficient use of water. Almost every analysis shows that Israel's economy would be stronger and the total value of output would be increased if water was transferred from agriculture to industrial or municipal uses (Lonergan and Brooks 1994). The opposite appears to be true in the West Bank and Gaza Strip; those economies would be stronger if water could be transferred into agriculture. Careful analysis would be needed to say whether other countries in the region lie closer to the Israeli or to the Palestinian case.

In conclusion, the social gains from approaching water problems from the demand side are very high and not restricted to direct financial savings. Reducing demand is also a very effective strategy (perhaps the most effective strategy) to minimize risk and to reduce environmental damage. The demand approach suggests that in most countries of the region, some water could be transferred from agriculture to other sectors, with overall gains for the standard of living and, very possibly, for the quality of life. The implication is that the long-term demand for water is much more elastic to price and policy than is recognized. In contrast, some analysts believe that water-short areas, such as the Jordan Valley countries, will have no alternative but to turn to external sources by early in the next century (Assaf et al. 1993). These analysts argue that even a higher level of end-use efficiency and a total shift of freshwater out of agriculture would be insufficient.

Water quality: the ecological crisis

Water quality, the second component of the regional water crisis, is less ancient but equally pressing. In the spring of 1994, five nations (Bahrain, Jordan, Lebanon, Syria, and the United Arab Emirates) participated in an environmental conference organized by the American University of Beirut, and each of them identified water pollution as a critical issue. The key point is that — again by analogy to energy — it is just as important to conserve the quality of water as to conserve its quantity (Brooks 1994; Lonergan and Brooks 1994). According to Assaf et al. (1993), Israel is losing 3–10 × 10^6 m^3/year of drinking water because of declining water quality.

Most water-quality problems derive from one or more of four factors: overpumping of aquifers, runoff from agriculture, discharge of human and industrial wastewater, and loss of habitat.

Overpumping of aquifers

Overpumping of wells causes a decline in the water table. During the recent drought, when aquifers were pumped particularly hard, water levels in Israel were falling typically by 10–40 cm/year. Unfortunately, overpumping, or "mining," of what should be renewable aquifers is all too common in this region.

A decline in the water table has several adverse effects. At a minimum, it adds to pumping costs and increases energy use. More important, a lower water table permits lower quality water to flow inward and contaminate the freshwater of the aquifer. Many of the countries in the region have coastal aquifers that, in their natural state, are 3–5 m above sea level; this, in turn, creates an outward pressure that blocks the inflow of seawater. Pumping, or, more accurately, overpumping, has lowered the freshwater level to below sea level, so the effect is reversed and salt water from the Mediterranean can now be found 1–3 km inland.

Runoff from agriculture

Irrigation is obviously good for farmers (MacLean and Voss, this volume). However, irrigation systems also pose environmental problems. In most countries in the region, agricultural runoff is the major non-point source of pollutants, including sediment, phosphorous, nitrogen, and pesticides. Per-hectare use of pesticides and fertilizers in Israel, Jordan, and Palestine rates among the highest in the world, and runoff is correspondingly high. One result of this is that over the past two decades, nitrate (from both fertilizers and reused sewage effluent) concentrations in the coastal aquifer underlying Israel and the Gaza Strip have doubled (Gabbay 1992). In Syria, Al-Sin Lake, the main freshwater source, is polluted by runoff. Such problems are anything but inevitable. Practices like conservation tillage, contour planting, terracing, and the use of filter systems can control soil erosion and reduce phosphorous and nitrogen runoff by up to 60% (World Resources Institute 1992).

Problems are magnified at the greenhouses and poultry factories, which are increasingly widespread in the region. Greenhouses are periodically rinsed, with as much as half of the chemicals going directly into the soil. Good practice isolates these operations from contact with groundwater and recycles the rinse water, but good practice is uncommon. In addition, the otherwise attractive use of brackish water for irrigation can increase soil salinity. Washing out the salts with freshwater can alleviate local problems, but this would allow the salts to drain into watercourses or aquifers, with potential long-term problems. For this reason, irrigation with brackish water is subject to special regulations where it is done just above sensitive parts of the coastal aquifer in Israel.

Discharge of human and industrial wastewater

Cities in this region are old — in some cases, ancient — and, just as with many newer cities, water-supply and sewer systems have either begun to deteriorate or cannot handle the growing loads placed on them. For example, the city of Jerusalem still discharges half its wastewater untreated into dry riverbeds. (A treatment plant is now being built.) In some cases, systems have been damaged by war. Water losses in Beirut went up from 40% to well over 60% during the 15 years of civil strife, and many sewage-treatment plans are inoperable because of shelling. Generally, however, urban areas in the region have adequate systems. In contrast, the situation is far from adequate in smaller cities and rural areas. In a few cases, as in much of the Gaza Strip, the need for investment in water-supply, drainage, and sanitation facilities is immediate.

It is difficult to assess the extent of industrial contamination in the region because so few tests are done, and when tests are done, the results are seldom disclosed. Spot checks by the Ministry of the Environment in Israel have found concentrations of specific contaminants at levels that are a few to 100 times the levels allowable in European countries. Conditions elsewhere are unlikely to be any better. Throughout the region, dumping of industrial wastes is common, sometimes directly into watercourses and sometimes into wadis, which, at the next rainfall, allows contaminants to seep into aquifers. Cleaning a polluted river is difficult; cleaning a polluted aquifer is much more so, and in some cases is simply not possible (Goldenberg and Melloul 1992). Even agricultural processing has its problems. Olive-oil mills, an otherwise excellent innovation that increases the value added from farming and provides employment in rural areas, produce both solid and liquid residues. The solid residues can be put back on fields, but the liquid residues have so high a biochemical oxygen demand that they are generally just dumped. The impact of about 40 mills in Jordan is equal to that of a city of 1 million people.

Loss of habitat

Finally, water quality in the region is being seriously degraded by losses of natural habitat, mainly wetlands. As a result of decisions to drain swamps, canalize rivers, or expand the agricultural frontier, water that was providing habitat for a multitude of plant and animal species is lost. Dredging and reclamation of land to expand urban space in Bahrain has not only destroyed commercial fishing grounds but also blocked natural drainage of agricultural land and increased the salinity of groundwater.

Why are these losses important? It is because water in place and the habitats it supports have value. Some of the values of in situ water, such as those associated with fisheries and hydropower or even with the prevention of subsidence above an aquifer, can be measured in conventional economic terms. Other values are partially calculable, such as those associated with recreation and tourism or with the dilution or purification of wastes. Some values are extremely difficult to capture in economic terms, like those associated with the regulation of river flows or the support of plant and animal habitat.

Losses in the region resulting from uncontrolled use of wetlands are unknown but clearly high. For example, the King Talal reservoir is too polluted for recreational use, but, as the only standing body of water in Jordan, this pollution carries an extraordinary opportunity cost. Wetland conversion can also be controversial. Construction of the Jonglei scheme to increase water flows to Egypt and Sudan "was stopped in the early 1980s as a result of violent opposition by the local communities who did not want their livelihoods and ways of life changed by the draining of the swamps of the Sudd" (Allen 1994). In the case of the Hula Swamp, the draining of which was Israel's first megaproject, plans are under way to restore part of the drained area to its original ecology.

Water equity: the political crisis

Water, not oil, has historically been at the heart of most political conflicts in the countries in this region. This section considers the internal institutions that have been developed to manage conflicts among sectors; and the international institutions that have been developed to manage conflicts among nations.

Internal institutions

This region is characterized by some of the largest and most sophisticated water-management agencies in the world. By and large, they have achieved the goals set out for them. They manage the water systems within their jurisdiction with great care, and they have developed impressive databases that permit control on a well-by-well or pump-by-pump basis. The real problems lie deeper — one begins to question the goals themselves and the structures erected to achieve these goals.

In every country of the region, water-management institutions are oriented to the goals of supply management (construction of dams, storage reservoirs, and other engineering works), with little attention to demand management. Further, the national institutions typically devote most of their attention to large-scale, centralized forms of supply management. Small-scale, decentralized options tend to be neglected or left to communities. The national institutions tend to be insensitive to indigenous practices, gender concerns, ethnic groups, and the environmental impacts of the institutions' actions. Such organizations merely reflect the concerns of the governments that created them.

Water-management agencies in this region differ only in degree from their counterparts in most other countries, North or South. Their true distinctiveness lies in two other characteristics: the centralization of water management at the national level and their close relationship with national agricultural agencies. Every one of the Middle Eastern countries has a ministry or senior agency in control of water affairs. Lebanon, for example, has the Ministry of Water and Electricity. Jordan has the Ministry of Water and Irrigation. In Israel, the Water Commissioner, a powerful official who controls planning, construction, and management of the nation's water system, reports to the even more powerful Minister of Agriculture. The situation is mixed in Syria and Egypt, where central agencies maintain control over irrigation water, and domestic water supply is left to local or municipal agencies.

The close political association of water and agriculture means that intersectoral conflicts tend either to be ignored or to be resolved in favour of farmers. It also means that internal water institutions resist suggestions to increase water prices for farmers or to move toward any form of water market or other means of establishing rational allocation. (There are many policy choices between volume allocation and pure markets that can provide for efficiency and equity.) As a result of the use of nonmarket prices in the face of limited water supplies, central control is required to impose allocations by volume or time of use; in rural areas, like those along the upper Nile, traditional patterns of allocation may still hold. Outside agriculture, prices are less closely controlled, and there is less need for allocation. In many countries (notably, Iraq, Jordan, Libya, Syria, and Yemen), demand is supply limited because of the infrastructure being unreliable or undersized or because of the poor quality of the water, particularly in the summer.

An exception to the bias in favour of agriculture occurs in times of drought. When water-supply allocations must be cut back, farmers typically bear the brunt of cutbacks. No sector can reduce water use so extensively and so quickly as agriculture.

International institutions

Surface water commonly crosses or forms an international border; aquifers commonly underlie a border. For a somewhat larger part of the Middle East, Rogers (1994) counted 25 international rivers. I know of no comparable tabulation for aquifers, but two examples are the Disi Aquifer, which underlies the border of Jordan and Saudi Arabia, and the Mountain (Yarkon–Taninim) Aquifer, which underlies Israel and Palestine. The Litani, in Lebanon, is one of the few rivers carrying more than 500×10^6 m^3/year that does not cross an international border.

International water is almost everywhere a subject of intense debate, with the discussion dominated by international lawyers and diplomats, rather than by social or physical scientists. In the Middle East, the basic principle for sharing water remains that of equitable use. This implies that the ways specific bodies of water are shared must be negotiated to fit the physical, economic, and social context of the parties involved. The rights of parties to specific quantities and qualities of water remain a contentious issue. In these circumstances, it might be helpful to shift attention from rights aimed at the supply side to rights to guarantee certain levels of demand. This is the effect of an approach supported by an Israeli–Palestinian team (Assaf et al. 1993), who proposed entitlements of 125 m^3 of potable water per person per year (Shuval 1992).

Although international law applies most directly to surface water, each of the principles used in dealing with surface water applies to underground water, qualified of course by the limited knowledge of aquifer hydraulics and the greater difficulty of monitoring. A model treaty for internationally shared aquifers has been drafted (Hayton and Utton 1989), but it has not yet been extensively discussed by politicians.

Discussions about international waters, including those in the Middle East, typically conclude with a call for basin-wide or aquifer-wide commissions to manage them as a unit. In my view, such schemes are visionary or, at best, premature. There is simply too little trust among these nations to consider joint management. It has taken the United States and Canada many years to establish joint procedures for management of the St. Lawrence River and almost as long for the Netherlands and Belgium to learn how to manage the aquifer that underlies their border.

This go-slow approach toward international management is not intended to preclude cooperation by way of prior notification of changes in river regime or specific joint institutions, such as those for research. Nor does it exclude the possibility of true joint management in those cases, such as the Mountain Aquifer, where Israelis and Palestinians really have no other alternative (Feitelson and Haddad 1994). Even in these cases, step-by-step movement toward cooperation with parallel but not united institutions on either side of the border would probably be more successful than attempts to move quickly to regional institutions.

Although joint management seems premature for quantity issues, it may not be so for quality issues. Competing demands for water rights have something of a zero-sum aspect about them, whereas environmental problems can affect all parties together. Nations that share water resources, particularly aquifers, should therefore experiment, at first, with joint water-quality management.

Militarization of water

Over the years, many people have argued that a war over water in the Middle East is more or less likely (most recently, Bulloch and Darwish 1993). It is true that, at times, shots have been fired and bombs dropped on water installations. Skirmishes were occurring between Israel and its neighbours just prior to the 1967 war, and Israel bombed a partially completed dam on the Yarmouk, late in that war. Iraq destroyed much of Kuwait's water-desalination capacity during the Gulf War. However, to go from these examples to a general proposition of water wars ignores the wide range of options available for overcoming water scarcity, such as drip irrigation and shifts to crops that consume less water. Such approaches can relieve the pressure much less expensively and with much less risk than military conflict.

Consider what is presumably the point of greatest dispute: the Jordan Valley. Even here, where the allocation of water is anything but equitable, water problems stem as much from internal economic decisions as from the special conditions of military occupation (Elmussa 1993). Using a different approach, researchers in the Harvard Middle East Water Project have reached the same conclusion. Depending on the particular resolution of the property rights to water, the total value of water in dispute between Israelis and Palestinians cannot exceed $600 million CAD per year (in 1996, 1.36 Canadian dollars [CAD] = 1 United States dollar [USD]) and probably lies closer to $200 million CAD. This is not very much money in international terms. The annual cost of water loss appears to be well under the daily cost of modern warfare.

If water wars in the Jordan Valley are unlikely, one wonders whether they will occur anywhere. There are simply better alternatives than war. However, these may not be politically easy or free of conflict. "Water shortages will aggravate tensions and unrest within societies," but, as opposed to outright warfare, "internal civil disorder, changes in regimes, political radicalization and instability" are the more likely consequences (Homer-Dixon et al. 1993).

Research as part of the solution

Research priorities should cover three categories: technical, socio- and enviro-economic, and institutional. The following are the most pressing research gaps.

Technical studies

1. Agricultural techniques appropriate for water scarcity:
   - use of poor-quality or saline water;
   - degree of natural recycling under different conditions; and
   - long-term effects of recycling irrigation water and treated wastewater.
2. Aquifer hydraulics and potentials:
   - discontinuous or karstic formations; and
   - fossil aquifers.
3. Alternative sources of supply:
   - rooftop harvesting for drinking water; and
   - rainwater harvesting and savanization for improved ecological conditions and farming.
4. Existing and alternative strategies in agriculture and industry for times of water stress.

Socioeconomic and enviroeconomic studies

1. Application of "soft energy" approaches to water to determine how far the analogy can be pursued and whether comparable policies could be proposed.

2. Careful estimation of the elasticity of long-term water demand to combinations of price, income, and policy change.

3. Better definition of noncommercial services, such as recreation; of environmental services, such as habitat preservation; and of water in situ.

4. Evaluation of market-based options for national or regional water management:

    - efficiency and equity effects of marginal-cost pricing and other pricing structures on various sectors, ecosystems, and classes;
    - efficiency and equity effects of alternative quasi-market allocation techniques;
    - methods for adjusting pricing for different qualities of water supply and of wastewater runoff; and
    - approaches based on international trading at nationally determined prices (such as the Harvard Middle East water model).

5. Review of traditional methods of augmenting water supply and limiting water demand to see how they compare (in efficiency, equity, and gender effects) with modern methods.

6. Evaluation, using various criteria, of the range of megaproject and regional import options for major increments of water supply, to develop a preferred ranking under various conditions.

Institutional studies

1. Better identification of the barriers to the adoption of, or investment in, water-saving technologies; and design of policies to lower those barriers.

2. Comparison of market and nonmarket institutions for distributing water efficiently and equitably.

3. Options for joint or shared management of transboundary water resources, particularly with respect to water quality.

4. Options for community- or common-property management for water.

5. Measures to increase awareness of the need and the means to conserve water.

6. Improved design for water utilities that incorporate

    - water supply and wastewater removal and reuse;

- supply-side and demand-side concerns; and
- economic, ecological, and social issues.

Conclusion

What the Middle East faces is not so much a water crisis as a chronic problem escalating to crisis dimensions because older problems are deepening at the same time as newer ones are becoming evident. With few exceptions, the countries in this region have already reached or are fast approaching the limits of their indigenous water supplies. In the absence of imports, greater efficiency in water use or shifts of water from one sector to another are the only options left, except for those few countries with enough energy to run desalination plants. Greater efficiency in water use may be encouraged by the existing institutions; shifts of water from one sector to another are almost never encouraged by the existing institutions.

Water is the consummate political issue in the Middle East. The role of research should be to ensure that economically efficient, ecologically sustainable, and politically acceptable alternatives are developed and put forward forcefully enough to lead to both the necessary internal adjustments and the equally necessary international negotiations. In the absence of political movement, water could, indeed, be a destabilizing or disruptive element in national and international relations.

References

Allen, T. 1994. Water: a substitutable resource? Department of Geography, School of Oriental and African Studies, University of London, London, UK. Unpublished paper.

Assaf, K.; al Khatib, N.; Kally, E. Shuval, H. 1993. A proposal for the development of a regional water master plan. Palestine Center for Research and Information, Jerusalem: Israel.

Bakour, Y.; Kolars, J. 1994. The Arab Mashrek: hydrologic history, problems and perspectives. *In* Rogers, P.; Lydon, P., ed., Water in the Arab world: perspectives and prognoses. Harvard University Press, Cambridge, MA.

Brooks, D.B. 1994. Economics, ecology and equity: lessons from the energy crisis in managing water shared by Israelis and Palestinians. *In* Isaac, J.; Shuval, H., ed., Water and peace in the Middle East. Elsevier Scientific B.V., Amsterdam, The Netherlands, pp. 441–450.

Brooks, D.B.; Peters, R. 1988. The potential for demand management in Canada. Science Council of Canada, Ottawa, ON.

Bulloch, J.; Darwish, A. 1993. Water wars: coming conflicts in the Middle East. Victor Gollancz, London, UK.

Elmussa, S. 1993. Dividing the common Israeli–Palestinian waters: an international water law approach. Journal of Palestinian Studies, 22, 57–77.

Falkenmark, M.; Lundqvist, J.; Widstrand, C. 1989. Macro-scale water scarcity requires micro-scale approaches. Natural Resources Forum, 13(4).

Feitelson, E.; Haddad, M., ed. 1994. Joint Management of Shared Aquifers: the first workshop. A cooperative research project. Palestine Consultancy Group, East Jerusalem, Israel; Truman Institute of The Hebrew University, Jerusalem, Israel.

Gabbay, S. 1992. The environment in Israel. National report to the 2nd United Nations Conference on Environment and Development. Ministry of the Environment, Jerusalem, Israel.

Goldenberg, L.C.; Melloul, L.C. 1992. Restoration of polluted groundwater: is it possible? Israel Environment Bulletin, 15(1), 16–24.

Hayton, R.D.; Utton, A.E. 1989. Transboundary groundwaters: the Bellagio draft treaty. Natural Resources Journal, 29, 663–722.

Hillel, D. 1994. Rivers of Eden: the struggle for water and the quest for peace in the Middle East. Oxford University Press, Oxford, UK.

Homer-Dixon, T.F.; Boutwell, J.; Rathjens, G. 1993. Environmental change and violent conflict. Scientific American, 268(2), 38–45.

Issar, A.S. 1994. Water under the deserts of the Middle East. *In* Isaac, J.; Shuval, H., ed., Water and peace in the Middle East. Elsevier Scientific B.V., Amsterdam, The Netherlands, pp. 351–362.

Kahana, Y. 1991. Water conservation measures in Israel. Israel Center for Waterworks Equipment, Report prepared for UNEP–Mediterranean Action Plan.

Lonergan, S.C.; Brooks, D.B. 1994. Watershed: the role of fresh water in the Israeli–Palestinian conflict. International Research Development Centre, Ottawa, ON.

Moore, D.; Seckle, D., ed. 1993. Water scarcity in developing countries: reconciling development and environmental protection. Winrock International Institute for Agricultural Development, Arlington, VA.

Rogers, P. 1994. The agenda for the next thirty years. *In* Rogers, P.; Lydon, P., ed., Water in the Arab world: perspectives and prognoses. Harvard University Press, Cambridge, MA.

Shuval, H. 1992. Approaches to resolving the water conflicts between Israel and her neighbours — a regional water for peace plan. Water International, 17, 133–143.

World Resources Institute. 1992. World resources 1992–93: a guide to the global environment. Oxford University Press, New York, NY.

SOURCES OF STRAIN AND ALTERNATIVES FOR RELIEF IN THE MOST STRESSED WATER SYSTEMS OF NORTH AFRICA

Mohammed S. Matoussi

Resource Economist, Faculty of Economic Science and Management, University of Tunis, Tunis, Tunisia

Introduction

North Africa, which includes the three countries of the Maghreb — Morocco, Algeria, and Tunisia — is one of the most water-scarce regions of the world. The demand for water from all sectors has increased tremendously because of rapid economic and population growth. The immediate result has been acute scarcity in some countries. If this growth in demand continues at its current rate, an overall deficit will be unavoidable.

The countries of the Maghreb, faced with the possibility of chronic water shortages in the near future, must increase supply or control demand. The options for increasing supply are very limited. The most accessible water sources have already been exploited; exploiting those that remain would require heavy investments and might cause significant, perhaps irreversible, environmental degradation. Even if the Maghreb succeeded in exploiting all of its potential resources, an overall deficit would appear in the near future.

This paper examines alternatives for reducing the chronic deficits that could be effected within the framework of the current economic situation. It also looks at possible mcasures for reducing demand after appropriate economic and institutional reforms have been instituted.

Overview

North Africa, extending across 2 000 km, is situated between the Atlantic Ocean and the Mediterranean Sea to the north and the Sahara desert to the south. The Maghreb is a land of contrasts. Parts of the region are green and lush, but others are very hot and dry. Rainfall is, of course, the major climatic factor. Commercial and subsistence agriculture depend significantly on the vagaries of the weather.

According to hydrologists, countries with less than 1 700 m^3 of freshwater per person per year will experience periodic shortages; below about 1 000 m^3, the shortages will be chronic. North Africa, with 611 m^3 of exploitable water per person in 1992, is therefore below the critical threshold of chronic shortages. Projections give an even more frightening picture: by 2020, renewable freshwater availability per person per year will fall to 378 m^3. Worse yet, these averages for the whole of the Maghreb mask enormous disparities among the three countries. Morocco is well supplied with water, but Tunisia is one of the most water-scare countries of the world, with 432 m^3/person per year, and projections indicate that this number will fall to 285 m^3 by 2020.

Water uses

The Maghreb countries receive an average rainfall of about 257×10^9 m^3/year. Only 37.2×10^9 m^3 is exploitable with existing technology. In 1991, 17×10^9 m^3, which included 10.1×10^9 m^3 of surface water and 6.9×10^9 m^3 of groundwater, was exploited. The index of exploitation (the ratio of exploited to potential resources) was 32% in 1992.

Irrigation consumes most of the water: 14.65×10^9 m^3/year. Morocco uses 10.2×10^9 m^3 of its water resources for irrigation (60% of the water in the Maghreb). Households and industry use 2.18×10^9 m^3/year (13%). This shows that the industrial sector is not very well developed and that the supply to households is inadequate. Table 1 shows the allocation of water to these sectors.

Table 1. Use of water in the Maghreb.

	Tunisia (%)	Algeria (%)	Morocco (%)	Total (%)
Households and industry	18.5	26	8	13
Irrigation	88.5	74	92	87

Households

Groundwater sources are generally used for potable water. Morocco gets 62% of its drinking water from such sources, and Tunisia gets 53%. There is a need to shift to surface sources because accessible aquifers are becoming depleted.

The reliability and quality of water vary from urban to rural areas and from one country to the other. Tunisia has the highest quality drinking water. The urban population is served by a public distribution network, and about one third of the rural population has access to safe drinking water. Major efforts are being made to improve the quality of the water supplied to the rest of the rural population. In Morocco, 75% of the urban population and 30% of the rural population are supplied with water from public systems. Algeria, on the other hand, has the greatest difficulty providing water to both its rural and its urban populations.

Irrigation

Irrigation is now used on about 1.6×10^6 ha, of which 716 000 ha is in public areas served by La grande hydraulique, a huge irrigation project. Morocco has the largest area (460 000 ha) under modern irrigation systems. It also has the largest irrigated area, with 1.3×10^6 ha (60%) of the 2.25×10^6 ha in the Maghreb.

Although irrigation has been practiced for centuries in many regions of North Africa (such as Rif, Dir of the Atlas, Haouz of Marrakesh, Cap-Bon, and Sahel), La grande hydraulique is the biggest agricultural development since independence. The project has tremendously increased the region's agricultural output, but not without some adverse effects:

- Excessive irrigation has led to waterlogging by raising the water table.
- Eutrophication and salinization are becoming common.
- Inappropriate management of watersheds has led to silting of reservoirs built at huge cost.

Water balance in the Maghreb

If current trends of consumption continue and there is no major structural change in the allocation of water or in the technology of exploitation, there will be water shortages throughout the subregion after 2000. Tunisia will experience the most severe deficit, but even Morocco, with water resources significantly greater than the average for the region, will also experience difficulties.

Morocco can expect to face a chronic water shortage in the future. Table 2 shows that by 2020, with the exception of Rifains du Nord and Sebou, every watershed will be affected. The large Atlantic west-central region, which is undergoing rapid economic development and urbanization, has major water resources, including several coastal streams and rivers, yet it faces a deficit to the tune of 850×10^6 m^3/year.

Projections indicate that Algeria will have an overall chronic deficit of about 245×10^6 m^3/year by 2010. The Algiers region, already experiencing declining quality of water from phreatic sources, aggravated by industrial, agricultural, and urban effluent, will have serious shortages by 2010.

Table 2. Water balance in Morocco projected to 2020 by region.

	Volume (× 10^6 m^3)		
	Supply	Demand	Net
Rifains du Nord	1 545	1 052	+493
Moulouya	1 670	1 816	−146
Sebou	4 768	3 916	+852
Bou Regreg et Côtiers Atlantique	852	902	−50
Oum Errabia	4 067	4 869	−802
Tensift	1 221	1 630	−409
Souss	777	1 175	−398
Massa	144	185	−41

Of the three countries of the Maghreb, Tunisia will face the most serious water deficits ion 2010. Projections suggest that the index of exploitation for water will exceed the renewable limit and go well beyond it, which means Tunisia will be mining its water system. No matter what scenario is chosen, Tunisia is faced with the most worrisome deficit.

Regions of chronic water shortage in the Maghreb

Although there are disparities in water availability among the three countries of the Maghreb, disparities within each country are even greater. This section briefly reviews some characteristic cases and suggest ways to alleviate the growing problems.

Morocco

The most water-scarce region of Morocco is Moulaya, which receives irregular rainfall. According to estimates. the annual water requirements there will soon reach 1.2 × 10^9 m^3, which will exceed available supply by 200 × 10^6 m^3, or one fifth of anticipated needs. Since independence this region has witnessed increased exploitation of water sources through intensive mechanization and deep pumping. The combined pressure from expanding agriculture, industry, and tourism, along with the effects of the droughts during 1972–86, has led to a steady lowering of the water table, exceeding 30 cm in some areas. No other part of Morocco has been affected as seriously as Moulaya. However, growing deficits in the region of Souss–Massa are expected to provide an equally serious challenge in the coming years.

Algeria

The Oran region, with an annual rainfall of less than 300 mm, faces severe water shortages, which limit large-scale irrigation and reduce industrial output. In addition, the water-distribution system is inadequate, and the basic sanitary standards for water are not respected. The aquifers of Mascara and Sidi Be Abbes provide more than 150×10^6 m^3 of water annually for irrigation but are being overpumped to the extent that degradation is almost inevitable.

Tunisia

Water shortages in Tunisia are chronic in the coastal region, which has the greater part of the population and major economic activities (large urban centres, factories, and hotels). Surface water is very limited along the coast and represents only 2% of the national potential. Deep aquifers are rather more widely distributed but still represent less than 10% of the national potential. Phreatic water alone is in major supply. Unfortunately, these shallow sheets of groundwater are threatened by pollution and overpumping.

Alternatives for reducing chronic deficits

To address the water-scarcity problem, the countries of the Maghreb must use existing technical means to increase the water supply.

To match water supply to demand, water-resource managers have the following options:

- treating wastewater;
- reducing distribution losses;
- improving drainage systems in agricultural areas;
- using brackish water; and
- developing a master plan for soil and water conservation.

If implemented, these approaches could increase supply dramatically. For example, Tunisia could recover up to 1.6×10^9 m^3 of water (Table 3), equivalent to half of the water resources currently being exploited by the country.

Table 3. Estimated annual volume of water recoverable in Tunisia.

Options [a]	Volume ($\times 10^6$)
Treating wastewater	200
Stemming distribution losses	700
Improving drainage systems	200
Conserving soil and water	430
Harvesting water	47
Total potential gain	1 577

[a] These options are slightly different from the ones in the text: water harvesting is added here, but brackish water is excluded. Were the latter included, the total gain would be even higher.

Treating wastewater

Treated wastewater constitutes a resource that is not only constantly available but also increasingly available, with the development of cities, tourism, and industry. Because of growing demand for water, on the one hand, and limited water resources, on the other, the countries of the Maghreb have little alternative but to recycle treated wastewater for use in irrigated agriculture. Eventually, treated wastewater is likely to be used in other sectors, as well.

Reuse of wastewater has a number of advantages:

- It provides a permanent source of water for the agricultural sector while protecting the environment.
- It reduces the use of groundwater, contributes to the recharge of aquifers, and prevents seawater from intruding into aquifers along the coast.
- It reduces the cost of production and use of inputs in agriculture to the extent that the organic content of the treated water covers the nutrient requirements of the majority of crops.

Reuse of treated wastewater in agriculture is thus a technique that adds to the value of the water resource while it protects the environment. All three countries in the Maghreb have developed programs to use wastewater.

Morocco

In 1982 about 217 $\times$ 10^6 m^3 of wastewater was produced in Morocco. This is expected to reach 725 $\times$ 10^6 m^3 by 2000 and will certainly exceed 900 $\times$ 10^6 m^3

in 2020. The office for water research and planning (Direction de la recherche et de la planification de l'eau) has an ambitious development program to reuse this currently wasted resource. However, up to now, implementation of the program has been very limited; treated wastewater is used on only 700 ha of land devoted to vegetables and orchards around the cities.

Algeria

More than 350×10^6 m^3 of wastewater was disposed of in Algeria in 1979; and 660×10^6 m^3 in 1985. Total wastewater is expected to rise to about 1.5×10^9 m^3 in 2010, but projections suggest the possibility of reusing about 600×10^6 m^3 in that same year.

Tunisia

Tunisia has fully accepted the treatment and reuse of wastewater as one of the principal ways to reduce the need to desalinate seawater. In 1992, Tunisia already had 28 treatment plants around the country. The volume of wastewater treated is about 88×10^6 m^3, of which about 14.6×10^6 m^3 is reused in agriculture. The country has also prepared an ambitious program for expanding wastewater use. Some 57 new treatment plants are expected to be built before 2000. The capacity for treatment will then be about 200×10^6 m^3/year, almost one third of the total phreatic water potential of the country. Agriculture will use 95% of the treated wastewater to irrigate 20 000 ha of land. Treated wastewater will therefore play a significant role in the near future.

Reducing distribution losses

The physical efficiency of a water-distribution network can be measured by the difference between the water that goes into the pipe and the water that comes out. A new pipeline system should not have losses of more than 10%. Losses in the distribution systems of the three countries of the Maghreb are much higher. In Algeria, for example, losses in the urban system are currently 50%. The immediate result is that the majority of residents have running water only intermittently, which, in effect, means that shortages are common. In Tunisia, losses in urban pipelines amounted to 85×10^6 m^3 in 1993, which represents a loss of 28%. The absolute loss is almost exactly equal to the capacity of the two dams at Béni-Metir and Kasseb.

These pipeline losses are greater in the irrigation-water network. It has been estimated that an improvement in physical efficiency of just 6% in these networks would be equivalent to saving 270×10^6 m^3 of the water resources behind the dams in Morocco.

Recovery of the water losses in the urban and irrigation distribution networks in Tunisia would save 700 × 10^6 m^3/year. This represents one quarter of all water currently exploited and exceeds the total of all phreatic sources in the country.

Improving drainage systems in agricultural areas

Poorly managed irrigation fields lead to waterlogging and consequent environmental degradation. The problems of drainage are in some cases exacerbated other factors, such as

- flat terrain with poorly aerated soils;
- intensive irrigation following a long drought, which occurs in the southern parts of all three countries; and
- intensive irrigation in vast, modern fields, such as those around Tadla in Morocco.

Many studies have shown that drainage systems and soil improvement on agricultural land could translate into significantly greater productivity:

- Draining the land would permit easier access to more distant parcels of farmland, so agricultural work could be accomplished under optimal conditions.
- A reduction in net use of water would mean an increase in the productivity per unit of irrigation water consumed.

In Tunisia, where irrigation is used on more than 300 000 ha of land and takes more than 2 × 10^9 m^3 of water per year, drained water could be recovered for a second round of use. Studies show that it would be feasible to recover about 200 × 10^6 m^3 of water. However, research on this subject is still in its initial stages, so any steps toward actual recovery must be taken with caution.

Using brackish water

Brackish water includes groundwater with a salt content that exceeds 4–5 g/L. All three countries have important reserves of brackish groundwater.

Using irrigation water with a higher level of salinity than acceptable is already a common practice. However, this practice entails serious risks over the long term because brackish water affects the aquifers and damages the soil.

In Tunisia, where brackish water is abundant, authorities have decided to exploit it. A brackish-water treatment plant in the southeastern part of the country is expected to begin operations very soon.

The decision to desalinate brackish water will require, over the next few years, a program built around three components:

- an inventory of brackish aquifers, for determining the quantity and quality of the water they contain;
- explicit integration of brackish water in planning the use of future agricultural water resources; and
- implementation of a research program aimed at determining the optimal patterns of use of brackish water in agriculture and industry.

Developing a master plan for soil and water conservation

In the Maghreb, which is characterized by uncertain rainfall and by heavy runoff when it rains, soil and water conservation are crucial. Particularly after prolonged dry periods, flows of water cause a major and characteristic form of erosion, with serious consequences for water resources:

- Hydraulic works, such as dams, suffer rapid silting and thus a loss in their valuable production capacity.
- Fertile layers of earth and sometimes even the vegetation that covers the soil can be lost.
- Recharge of aquifers is insufficient.

If the three countries of the Maghreb wish to make major improvements in their water balances, they must develop a master plan for soil and water conservation, with a range of objectives, each integrated with the others. Among many possible objectives, the following seem to be the most urgent:

- reforestation in the watersheds behind major reservoirs;
- new small dikes and dams to retain water in the hills and mountains;
- a program to recharge near-surface aquifers and groundwater;
- a program to identify and evaluate favourable sites for water harvesting and spreading; and
- an intensive research and testing program to evaluate additional resources (including artificial recharge) and to optimize conjunctive management of surface water and groundwater.

All three countries of the Maghreb have already launched activities to promote soil and water conservation. Morocco has created an integrated management plan for surface and groundwater on the plain of Souss–Massa, which could serve as an example for other countries. Tunisia has established a 10-year program for the construction of 1 000 small lakes in the hills and 200 mountain dams, with the objective of recovering 160×10^6 m^3 of water per year.

Measures for reducing demand

The three countries of the Maghreb must design economic and institutional measures for the rational use of water. These measures would significantly improve water availability in the region. However, results will remain quite short of the potential if the economic and institutional framework remains unchanged.

There is need to shift toward measures that limit demand, because development of additional conventional supply is becoming more and more difficult. These countries have already exploited 43% of their water resources, those most accessible and least costly. Development of the remaining water resources will be even more complex technologically, and future investment will certainly be less profitable.

On the demand side, several measures can be implemented to create the needed change in water strategies. The measures appropriate to the current conditions in the Maghreb are the following:

- setting realist water prices;
- establishing a new strategy for irrigation;
- increasing public awareness and encouraging research; and
- integrating and decentralizing management.

Setting realistic prices for water

Any action to control demand, if it is to be effective, has to begin with modification of the price system for water, along with other means for allocating the resource. Pricing policies currently in effect (water tariffs) throughout the world, in humid countries as well as in arid, do not take into account the cost of bringing the water to consumers and neglect totally the opportunity cost of the water. Opportunity cost would be zero only if the supply of water were infinite.

Water prices are generally very much below real costs of delivery, and, in the agricultural sector, irrigation water is commonly priced below operation and maintenance costs. The immediate results are waste of the resource and poor service to users. However, policies are changing. The need to protect water resources, the increasing costs of supply and distribution systems, and the demand for adequate supply of water for domestic use and agriculture are some of the reasons for renewed interest in pricing systems for water as a notably effective means to the optimal management of the resource.

Water-pricing policies must be designed and implemented in the near future, and they should not only cover the direct operation and maintenance costs of supply and distribution and of management agencies but also cover, at least partially, the capital costs of the investment. The same policies aimed at covering costs will also reduce demand for water and promote its more rational use. The Maghreb countries have all adopted progressive pricing systems for drinking water, and these tariff systems serve as an illustration of what can be done. However, it is important to note that an optimal price system for water does not at all imply an excessive price increase. The exorbitant prices that are sometimes levied can be counterproductive if they induce big consumers to turn to other sources of supply.

Establishing a new strategy for irrigation

The countries of the Maghreb, especially Morocco, with its Million Hectare Program, have initiated very ambitious programs to expand irrigation. Indeed, modern irrigation has transformed whole regions, and it has enabled the Maghreb to increase its food production tremendously.

The competition for water for different sectors will invariably lead to its reallocation to those activities that are most profitable, at the expense of traditional and subsistence activities. To minimize the adverse consequences of this inevitable restructuring, an appropriate strategy is needed to establish a new role for irrigation that aims at optimal allocation. Some of the elements of this new irrigation strategy could be described as follows:

- Experience in the region has shown that private irrigation is generally both more profitable and more frugal in its use of water than large, public irrigation schemes. The new strategy should take this into account and reorganize the public irrigation schemes, at least by offering them more organizational and financial autonomy so that they may collect tariffs and provide services more effectively.
- Users should participate in the development and maintenance of their irrigation networks through specially created irrigation associations.

Increasing public awareness and encouraging research

One initiative that can contribute significantly to water conservation is the development of programs to disseminate information, to increase public awareness, and to encourage training and research. Among other things, such activities should include

- training senior and operational personnel for work in the field of resources conservation;
- training maintenance staff; and
- educating and training users.

The last of these is particularly important because water users have to be fully aware of the debt that their country will eventually have to face for its water supply.

Integrating and decentralizing management

Two major characteristics of water management in the Maghreb — and in most other parts of the world — are (1) the absence of integrated but decentralized management and (2) public monopolies that produce and distribute water without attention to basic criteria of profitability. These factors have led to inefficient allocation of water.

Integrated and decentralized management systems

The supply and distribution of water in the Maghreb are organized so that each use (irrigation, potable water, etc.) is managed by a central office following very rigid rules and operating independently of other offices. For example, in Morocco, irrigation is controlled by offices for agricultural development (offices de mises en valeurs agricoles); in Tunisia, by regional commissions for agricultural development (commissariats régionaux au développement agricole). These agencies are highly centralized and pay little attention to profitability. Generally, the supply of potable and industrial water is managed by public agencies that have social, regional, and even political objectives, whereas pipelines and water transportation are organized by independent public monopolies.

This fragmentation of water management leads to wastage and overemployment in the water sector. A single, conceptually global management system is necessary. This would strengthen linkages among alternative uses so that an efficient allocation of water would be attained.

The concept of a global and integrated system of water resources assumes that there is some preexisting definition of an appropriate spatial framework. The concept of such a system should include consideration of all externalities in its objectives. The watershed for example, has just these appropriate characteristics for all watercourses. Indeed, the natural or ecological interdependencies, sometimes called technical interdependencies, between water resources and agricultural resources are so linked that any planning in one sector will always affect the other.

Decentralized management agencies

Decentralization should be adopted in stages but, in the end, should be as extensive as possible. In the first stage, financial autonomy should be given to user organizations and should involve delivery of services at the full cost of the water. Where possible, decisions, such as those relating to repairs, maintenance tasks, and construction of sewage-pipe systems, should be left to the market.

Decentralization should also involve the creation of user associations, permitting them to progressively take over the management of certain components of the water system, such as irrigation networks.

Water markets

The ideal form of decentralization of water management would involve the creation and operation of water markets to allocate and manage water resources. The concept of water as a good involves a number of characteristics related to both its physical and its economic nature, and this explains why water supply is controlled or regulated by public bodies. However, recent experiments with water markets, mainly in the United States, have given encouraging results and have therefore raised interest in applying this method to water management in the Maghreb.

Nonetheless, if market forces are to determine access to water resources, some serious problems will ensue. Water, particularly groundwater, is common property, as well as being a public good. Therefore, purely private management would certainly lead to speculation and monopoly and result in waste of the resource. Low-income people, particularly those living in rural areas of developing countries, would be severely affected, as would those relying on subsistence and traditional agriculture. If a water market is to be feasible and work well, that is, with optimal allocation to defer recourse to nonconventional sources, it must have well-defined rules and mechanisms.

Water Crises and Constraints in West and Central Africa: The Case of Côte d'Ivoire

Jean Biémi

Lecturer, Faculty of Science and Technics, University of Abidjan, Abidjan, Côte d'Ivoire

Water resources

The main sources of water in West Africa are rain, surface water, and groundwater. In recent times, changes in climate, deforestation, and drought have seriously affected these resources in most parts of the subregion.

Rainfall

Rainfall diminishes progressively from south to north and as altitude increases. Some regions receive plenty of rainfall, and others get very little. Generally, however, much of the region receives sufficient rain, but the full potential of the rainy seasons has not been exploited. During certain months of the year, water is plentiful and sufficient to cover the dry spells that follow, but techniques to harvest this water are lacking.

The equatorial forests get plenty of rain, but their soils are poor. The peasants who live in the forests cannot afford fertilizers, so they keep clearing one patch of land after the other. This constant relocation makes the construction of waterworks impractical.

There is inadequate rainfall in tropical areas, so people rely on permanent rivers or wells. In tropical dry areas, rains are insufficient during the major part of the year, although the soils are rich. People use water from wells and boreholes to irrigate their farms.

The arid areas receive hardly any rainfall. They are occupied by nomads (with their herds), who live around grazing lands and oases.

Surface water

In West Africa, especially the humid zone, water is obtained from springs, marshes, lagoons, lakes, and national and international rivers.

Springs and streams

The humid zone has many natural springs and small streams. Springs are found mainly close to the shore and around mountains, such as in the Fouta Djalon in the Man region of Côte d'Ivoire. Some of these perennial sources are believed to have thermo-mineral qualities. Natural springs and streams, especially in Precambrian rocks, generally have excellent-quality water and have been used to supply potable water from time immemorial. Unfortunately, these springs are no longer used.

In 1964, most villagers around the Dans Hills in Côte d'Ivoire were relocated by the government. New villages were created after 1973, but the people dug wells, neglecting the springs.

Rational use of springs, especially in the mountainous areas and along the shores of West Africa, could play an important role in the supply of potable water to the region's growing population.

Marshes

In the sub-Sahelian zones, natural springs are scarce, and most of the watercourses have seasonal flow. There is an acute shortage of water, so villagers fetch water from marshes. Even where there are wells, water from marshes is used for household activities (washing, building, etc.), to conserve the wells' limited capacity. In some cases, marsh water is also used for drinking, either because the water from the wells is insufficient or because people do not like it for some reason (unusual taste, or location in a sacred forest or near a cemetery). Because water from the marshes is of poor quality, waterborne illnesses, such as diarrhea, dysentery, and cholera, are common. Endemic goitre has for a long time affected people who live around Danané and Biankouma (western region).

Lagoons

Côte d'Ivoire is well endowed with lagoons, which include Aquien, Aby, Ebrié, Tendo, Potou, and Egny. Local people are proud of these lagoons, and Abidjan is commonly known as the Pearl of Lagoons. The lagoons have various uses: sailing, boaters' dances, pirogue (canoe) races, swimming, fishing, and international trade and transportation. The lagoons also play a role in religious life: parts of a lagoon may be worshipped, and sacrifices may be offered to the water spirits on the eve of a fishing trip.

Unfortunately, the lagoons are easily polluted. High rainfall (Abidjan gets 2 800 mm/year) washes away the soil, carrying it into the lagoons. However, pollution is mainly caused by human activities: most of the city's domestic and industrial wastes end up in the lagoons. The problem is aggravated by the pressure and waterproofing effects on a sandy soil from the weight of buildings, which forces water down from the surface, either vertically toward the deeper sheets or laterally toward the lagoons.

Freshwater sources are further threatened by the opening of the lagoons to the sea. Before the construction of the Vridi canal, lagoon water was fresh, but today the lagoons receive 10 000 m^3 of seawater annually, which nearly equals the 12 000 m^3 of freshwater they receive from continental rivers every year.

Lakes

The two main lakes in the subregion are Lake Chad, along the borders of Niger, Nigeria, Cameroon, and Chad, and Lake Nyos, which lies within Cameroon and Nigeria. Conflicts often occur among the countries sharing Lake Chad, when, for example, the level of the lake declines and small islands emerge, which are quickly invaded by fisherfolk from each country. More than 50 agreements have been signed by Cameroon, Nigeria, and the Central African Republic to regulate the exploitation of the Lake Chad basin.

National rivers

In the humid zones of West Africa, national rivers are the main sources of water supply for urban centres. Major construction, such as dams, affect not only the ecosystem of the watercourse but also local people, who derive no benefit from either the electricity or fishing. They are often relocated and end up losing their livelihoods. For instance, the Baoulés, from the central part of the country, were relocated to the southwestern forests in San-Pédro, among the Krou ethnic group. This has posed some serious problems because the mixing of various ethnic groups has often led to fights over land, sometimes culminating in deaths.

Rivers are often polluted, which is a particularly serious problem where they are the sources of potable water. For example, in the main cities of Cameroon, Yaounde, and Douala, the government has to use expensive chemicals to treat water for urban residents.

International rivers

The five international rivers in West Africa are the Senegal River, which originates in Fouta Djalon in Guinea, crosses Mali, and enters Senegal after crossing several kilometres along its boundary with Mauritania. The Niger River also originates in Fouta Djalon; it flows through Guinea, Burkina Faso, and Nigeria, before it empties into the sea. The Bénoué, a tributary of the Niger River, originates in northern Cameroon, before joining the main course in Nigeria. It links Guinea, Mali, Côte d'Ivoire, Niger, Burkina Faso, Nigeria, and Cameroon.

The Volta originates in Burkina Faso and empties into the sea in Ghana. Its main three tributaries are Black Volta, White Volta, and Oti. The Logone River originates in Cameroon and the Central African Republic and empties into Lake Chad, and the Chari flows from the Central African Republic to Lake Chad. These two watercourses link Cameroon, the Central African Republic, and Chad.

Conflicts can arise among riparian states over use of international rivers. Congo and Zaire often meet to talk about their common river, which has two names: Congo on one side of the border and Zaire on the other side. There are conventions regarding its use, although each country is keen to keep its name on its portion of the river.

Groundwater resources

Groundwater is found in loose soils near the shores, in alterites on crystalline bedrock, and in fissures in bedrock in the humid zones of West Africa.

Loose soils

In the sedimentary basins of the shores of Côte d'Ivoire, Togo, Nigeria, Guinea, Cameroon, and Sierra Leone, large volumes of underground water are found in the aquiferous strata of Quaternary sediments at a depth of less than 10 m or in older rocks at depths of 10–80 m. Additional resources are found in the Upper Cretaceous sediments, but only at 100–200 m depth.

In Côte d'Ivoire, phreatic sheets with discharges of 2–22 m^3/h are found in certain Quaternary sands, within the top 10 m. In Abidjan, water is drawn from these sheets at a depth of 3–4 m, from pits dug in the sand, and is used for washing hands and cars, raising pigs and poultry, making bricks, and watering gardens and flowers.

Water sheets of Miocene–Pliocene age are found at the continental edge in clayey river sands near the surface and in coarse sands at greater depths. Flow rates are 7–338 m^3/h, but depths may reach 200 m. Abidjan is better located for potable-water supply than other cities of West Africa. Although all the phreatic waters of the continental edge risk pollution from seawater, in Abidjan the beds feeding the city are sheltered from seawater incursions by the lagoon fault.

Beds of alterites on crystalline bedrock

Peasants and drillers are familiar with groundwater occurring in beds of alterites lying on top of crystalline bedrock. For a long time, they were considered the only sources of exploitable water resources on bedrock. These alteritic reservoirs reach 50 m thick in Côte d'Ivoire (100 m in the "cocoa loop" above volcanic sediments) and 10–20 m thick in the humid zones of West Africa. Deeper layers are composed of grainy grits and are the most productive and sought after. However, these zones are generally not more than 5 m thick. Exploitation of alteritic reservoirs is decreasing because of the depth at which the water is located and the thinness of saturated zones. Large-diameter water wells, even those of modern design, cannot normally reach beyond 30 m.

Alterite water reservoirs are subject to significant seasonal variation (decrease of the sheets in the dry season and refilling in the rainy season) and can easily be depleted because of the limited height of the water column. Local use by people with ropes and buckets can reduce its level. In alteritic beds in the north of Côte d'Ivoire, for example, 85% of the wells are characterized by a decrease of water level; only 4% show no decrease; and the remaining 11% show a slight increase.

The water of the alterites is used in the rural areas and some urban settlements that are not connected to the mains. Because irrigation is not well developed in West Africa, no real conflict has arisen over water coming from alterite sheets. The real problems stem from pollution and depletion.

Fissures in crystalline bedrock

Water in fissures of crystalline bedrock is generally of good quality and is used for human consumption. Some water is found in schists that have been folded and deformed, which can increase porosity. However, the greatest water potential occurs where the rocks are fractured. Original permeability is generally low, except in detritic layers, which play the roles of drains and locally develop important capacity, but at depth.

Migmatic reservoirs are aquiferous, which can be seen by the impressive number of water traces recorded in the bedrock. In fact, at depths of 120 m in Mali, 124 m in Côte d'Ivoire, and 400 m in the Tarkwa mines in Ghana, numerous traces of water were recorded, revealing the high water content of the undisturbed crystalline formations. Water circulates in those crystalline rocks through networks of fractures that serve as drains and that are capable of discharging sufficient quantities of water when reached by drilling.

As reservoirs, fissures appear to be more regular and more stable than the alterites. Fluctuations of water levels do not generally exceed 5 m one way or the other, which indicates rapid replenishment. Even after significant withdrawals, the wells readjust their water level as soon as there is a break in the pumping.

Exploitation of groundwater

There are various methods of exploiting groundwater. Sumps or ponds provide small quantities of water to some villages in the sub-Sahelian zone. These sumps are small holes up to about 2 m deep, dug into the alluvial soils near beds of small lakes or rivers. They often last for a very short time, from a few days to a week or a month, but never from one season to another. During the rainy season these sumps often overflow. Although large volumes of water can be drawn easily into these pits, the water is of very poor quality.

Peasant wells are typically 3–10 m deep and 1 m in diameter. Women draw water from the wells by hand. The wells are dug into surficial alterites on plateaus and hills close to villages. Rains easily feed these wells, but during the dry season, water levels decrease, and in some cases the wells dry up. These wells are dug by hand by Malian or Burkinabés sinkers, seldom by Ivoirians. The wells are tubeless, and the concrete coping, built 0.5–1 m above the soil surface, is the only means of protecting them from pollution.

Modern wells generally penetrate the whole thickness of the disturbed geological layer above the sound bedrock. However, the depth of the wells is limited to the sandy or clayey alterites and to the maximum capability of the drilling equipment (35 m). These water points are built with a diameter of 1–2 m (cistern wells), which allows them to store considerable amounts of water. This water is easily drawn by groups of women, each holding a rope attached to a pail. Maintenance costs are low, and breakdowns are rare. However, because of their limited depth (8–35 m), they are subject to the same seasonal effects as the peasant wells and generally do not reach the most productive levels (coarse-grained sands) at the top of the bedrock. Furthermore, the risk of pollution is high because of the possibility of seepage.

The diameter of deep wells is small, not more than 20–23 cm through alterites and 15–16 cm in bedrock. However, their depth reaches 124 m in Côte d'Ivoire (drilling of Brou Akpaoussou in Bondoukou). The protection of the wells is easy because they essentially gather water stored in the fractures of the crystalline bedrock. In Côte d'Ivoire, several types of pumping equipment are used, of which the most modern can exploit water located at depths of 80 m or more and discharge about 1 m^3/h.

Unfortunately, these pumps can be difficult to handle and often break down. Field surveys from 1984 to 1987 showed that of 212 wells constructed since 1982, nearly 23% were out of order at any given time. More than 8% of these wells were rejected by the population because the water had an unusual taste, or the wells were on sacred land or too far. Some 18% were dry. Estimates in 1986 indicated that 15–20% of the pumps were experiencing technical problems or breakdowns. An earlier investigation by the World Health Organization, in 1976, showed that breakdowns increased after three years of pump operations and that 40–80% of the pumps originally installed may no longer be working.

Water institutions and pricing

International organizations

A number of African countries have come together to create the Union africaine des distributeurs d'eau (UADE, African Union of Water Distributors). Member countries are Benin, Cameroon, Côte d'Ivoire, the Central African Republic, Congo, Djibouti, Gabon, Ghana, Guinea, Burkina Faso, Liberia, Morocco, Mauritania, Mali, Niger, Rwanda, Senegal, Sudan, Tunisia, Chad, Togo, and Zaire. Thanks to UADE, certain principles for improved water exploitation, distribution, and pricing are practiced by the member states. Perhaps it is even more important that West African states have created a water research organization, Comité interétat d'étude hydraulique, which has, among other things, improved the state of knowledge concerning fissured rocks bearing water.

Organization in Côte d'Ivoire

According to estimates, in 1964 more than 70% of the villages in Côte d'Ivoire lacked sufficient or good-quality water. Water-related diseases were endemic. From 1973, because of disparities in water distribution and a severe drought, the government launched a big water program: Program national de l'hydraulique. This program was supported by a national water fund, Fonds national de l'hydraulique, inside the Caisse autonome d'amortissement, and by an autonomous water service department in the ministry.

From 1973 to 1988, the plan aimed to

- supply potable water to all urban centres with more than 4 000 inhabitants; and
- equip villages with more than 100 people with water points, with supplementary points for each additional 600 inhabitants.

The program was highly successful. More than 270 communities were provided with potable-water systems, and more than 12 000 water works were built in almost 8 000 villages. By 1985, more than 80% of the population was supplied with potable water in both urban and rural areas.

Up to 1988, the national water program had invested 104 billion XAF (in 1996, 277 CFA Francs [XAF] = 1 United States dollar [USD]) for urban activities alone, plus another 40 billion XAF for water works in rural villages. Of this 144 billion XAF, nearly 9 billion was from the government and the rest was from international development agencies. In 1988, the program was revised, with the following objectives:

- provide water in 168 small urban and rural centres with more than 4 000 inhabitants;
- maintain the level of water supply, through programs to strengthen production utilities and extend distribution networks;
- restore existing waterworks in villages where they were exhausted, defective, or insufficient;
- increase the number of water points to meet identified needs, with a total of 15 000 to be installed by 1990; and
- create awareness and train villagers to undertake the maintenance of the pumps.

Water tariffs

In Côte d'Ivoire, water is treated as a commodity and sold at a price that reflects cost. Of course, in the sub-Sahelian areas, where water is scarce, cost of supply exceeds by far that in forest areas and especially that in Abidjan. Water in the city is subsidized, although it should be sold at a higher price than in those parts of the country that are close to sources, where piping is not needed.

The aim of water tariffs is to meet the costs of production, including equipment, installation, distribution, and maintenance. The full price is made up of three components: maximum basic price, surcharge, and development charges.

The maximum basic price corresponds to the different charges that are included in the production of 1 m^3 of water and a 5% profit margin for Société de distribution de l'eau en Côte d'Ivoire (SODECI, Society for the distribution of water in Côte d'Ivoire), a private company that manages water distribution.

The installations used by SODECI belong to the government. The government borrowed money for these water projects, and the surcharge added to the maximum basic price paid by the consumer is devoted to the loan repayment. The surcharges are kept in a special fund, which, until 1992, was managed by SODECI on behalf of the state. From 1992, the management of the fund was entrusted to the Caisse autonome d'amortissement.

The development charges cover recurrent costs: extension of pipelines, rehabilitation of equipment, modernization, and reinforcement of lines and other works.

The price depends on consumption, which, among other things, reduces the impact of water pricing on the incomes of small consumers. The four brackets include social (small-scale consumers), domestic (below-scale average consumers), normal (average-scale consumers), and industrial (big-scale consumers). In addition, there is a special rate for administrative use of water, which is payable as a lump sum. For consumers who pump directly, the price is currently set at 193 XAF/m^3, of which 188 XAF goes to the state and 5 XAF goes to SODECI.

Water is sold at prices ranging from 159 to 350 XAF m^3. The prices for this are about 200 XAF in Chad, 168 XAF in Congo, 120 XAF in Burkina Faso, and about 680 – 900 XAF in Guinea. In Cameroon, water is free up to a certain volume, and tariffs are applied only beyond this point.

A special problem of pricing occurs in rural villages where a well has been drilled but there is no pump. In such cases, there is no direct cost to provide water, but the cost of the pump must be amortized and payments must be made for well maintenance. The state requests some contribution from the villagers, but the rate is variously set: 10 XAF/pail of water in some cases; 25 or 30 XAF/family each week in other cases; and an annual fee of 1 000 XAF/person or 55 000 – 60 000 XAF per village for each pump in still others. Despite these charges, the state bears an annual cost of 100 million XAF for unreimbursed expenses, just to repair the pumps, and up to now no attempt to increase villager contributions has been successful.

Coordinated management and protection of water resources

Water legislation

Each country in Africa needs a permanent water code to define the techniques and conditions for water exploitation, distribution, and pricing, as well as the techniques and conditions necessary for collective and individual water use and sanitation. The code must include measures to prevent waste and pollution.

In Côte d'Ivoire, only the Abidjan water-map sheet was completed during the 1980s, and this sheet covered only one part of the aquifer. In the interior of the country, no town has a study on underground water. Consequently, in some localities, wells supplying drinking water and sanitary latrines are built at random, without taking into account the direction of the flow and the permeability of the soil. Consequently, contamination occurs.

A water code is necessary to define the African standard for potable water. To date, the continent does not have any specific standards for water. It uses the international ones, which do not take into consideration African particularities. For example, in rural areas, the nitrate content is often excessive in consumption water, without the population being aware of this.

The water code must provide for the creation of national and international commissions on management, protection, and use of all important water reserves. Agreements should be concluded on the water reserves to define the conditions of use by members. Such commissions should make special allowance for the particular interests of vulnerable populations. For example, a commission for the protection of the Ebrié lagoon, of Abidjan, could compel the counties to treat their wastes before pouring them into the lagoon, which would greatly improve conditions for the low-income people living on its shores.

Improvement of water institutions

In West Africa, each state has a company dealing with the exploitation and distribution of water: SODECI in Côte d'Ivoire, SNEC in Cameroon, SONEES in Senegal, SNE in Niger, etc. These companies formed UADE. For research, there is an Inter-African committee for water research (Comité interafricaine d'études hydrauliques), based in Ouagadougou. This committee deals with extension work to spread new technologies and information concerning water in Africa. As well, in the Sahel, an interstate committee for drought control (Comité interétat de lutte contre la sécheresse au Sahel) has been established. This committee is also based in Ouagadougou.

Results obtained by these organizations are already encouraging, but much remains to be done. In certain cases, the quality of service is below standard because of the heavy-handedness of the state in the management and control of water services. Maintenance is badly performed. Much water is wasted through leaks in the pipes, and leaks may last for months. Some people do not pay their water bills, and the loss must be made up by the small consumers, who are helpless and have no procedure for complaints.

It would be judicious to liberalize the field of water and let private societies play a greater role in exploiting water and distributing it in urban areas. This would create some competition and would lead inevitably to an improvement in the quality of service.

Water conservation policy

The water sector needs to be organized in most states of West Africa because there are many large losses of water and other forms of wastage, particularly in public services. However, consumers are unorganized.

Proposals for research

Social research

Displaced populations

Study is needed of the populations displaced by the construction of big dams. Follow-up is necessary to determine whether these people could return to their original lands or be compensated.

Marsh dwellers

Certain groups remain linked by tradition and by preference to their marshes, even where well water is clearly potable. It would be appropriate to study these people to see whether they could become interested in alternatives to marsh water without modifying their ancestral diet and other practices.

Institutional and economic research

Creation of databases

Attempts are being made in Africa to create a database to include both climatic parameters and surface waters. However, little has been done to gather information on groundwater, not even in those villages where wells have already been drilled and water prospecting is carried out. The necessary studies include the preparation of hydrogeological maps.

Management of pumps

Study is needed to develop regulations for village water systems to check the use of pumps that easily break down or are difficult to operate. For example, the SEEE pump widely used in Côte d'Ivoire needs at least two strong women on each handle to operate it. In other cases, pumps are very difficult to repair or require imported parts.

New options are needed for village waterworks to replace the present pumping system with a solar system or an electrical system using submersible pumps. In this case, distribution points may be established in the villages and maintained by selected people. The water could then be sold by the pail or other conventional container.

Reduction of waste

Management of water resources and improvement in distribution networks have to be studied to control waste. Leaks in the distribution facilities must be located and repaired.

Climate change in the Fouta Djalon

One of the most urgent research tasks in Côte d'Ivoire is to find ways to manage and reduce environmental vulnerability in the extension of the Fouta Djalon in the western part of the country. Even in their severely disturbed state, the forests of Man are the only source of wood for heating, construction, carpentry, and handicraft (tool handles, kitchen utensils, basketwork).

Humid-zone research

Coordinated management of the water–vegetation–habitat linkages is needed in the humid zones of West Africa. Lack of surface-water resources is more and more becoming a problem in certain regions, such as western Côte d'Ivoire, where vegetation is disturbed by the destruction of forests.

Man-Danané region

The region of Man-Danané has serious problems of demographic instability (refugees from Liberia), problems in agriculture (erosion of slopes and lack of land), difficulties with transportation (unevenness of soil surface), and climatic problems (degradation of forests). Research in this region would be important for the knowledge of water constraints resulting from the operation of natural systems.

Technical research

Aquifer rehabilitation and management

Research is needed on ways to reclaim and rehabilitate aquifers and on ways to manage water and treat runoff and wastewater to protect the underlying aquifers. Such projects must take into account the problems of waste management, agriculture, and the potential for seawater incursion.

Lagoon protection

Lagoons, especially Ebrié, in Abidjan, should be studied to develop a comprehensive protection plan, including the installation of treatment plants at the head of every sewer.

Rainwater use

Rains are very abundant in the humid zone of West Africa, but rainwater is not used except in the natural runoff across the countryside and the recharging of aquifers. Stored rainwater could serve many purposes, which would leave the wells to supply potable water.

Cistern design

In West Africa, water is abundant in the rainy season and scarce in the dry season. Simple designs and construction techniques for underground cisterns are required, particularly for villages located in the Sahelian and sub-Sahelian regions. Basins or excavations could store enough water in the rainy season to meet the continuing needs of the dry season.

Hydrogeology

Hydrological study is needed for all programs of village water development in crystalline rocks. The goal should be to create a map, or feasibility study, for each village. This would involve the analysis of aerial photography, remote sensing, and geophysics (electric probing and electric tracking). Hydrogeological studies are also needed in urban areas; fortunately, some are under way.

Effects of climate change

In many regions of Africa, there is consensus that climate has been changing for about 30 years: rains are becoming scarce; deforestation is common; and drought is more frequent. Study is needed to determine the extent of the apparent climate change and of its consequences for the water resources in Africa. In particular, we must study the vulnerability of water resources.

Strain, Social and Environmental Consequences, and Water Management in the Most Stressed Water Systems in Africa

George Khroda

Director, Research and Programs, REDPLAN Consultants, and Consultant for the Royal Netherlands Embassy, Nairobi, Kenya

Introduction

Mean rainfall in Africa is about 670 mm, but there are great regional disparities. Although Africa has abundant water resources, they are spatially and temporally maldistributed because of variability of climate, topography, and geology. Of these factors, climate is probably the most important because the distribution of rainfall is determined by the wind system, topography, and pressure of large water bodies.

The heaviest rainfall occurs near the equator, especially in the region from the Niger Delta to the Zaire River basin and central Zaire. Along the coasts of Sierra Leone, Liberia, and Madagascar, annual total rainfall exceeds 2 000 mm. Northward, rainfall decreases rapidly to about 250 mm at 18°N, except along the Mediterranean coast of Morocco, Algeria, and Tunisia, where the annual total ranges between 250 and 1 000 mm.

Rainfall distribution in the southern hemisphere is complex. Between the equator and the tropic of Capricorn, there is a north–south decrease, whereas farther south there is an east–west decrease (Fig. 1). Reliability also decreases, with variability of more than 40% in the deserts to less than 15% over the tropics. The humid tropics generally receive a lot of rainfall throughout the year, whereas the subtropical semi-arid regions experience marked rainfall seasonality and frequent droughts. The deserts are dry, and water is deficient throughout the year.

The semi-arid region is most affected by droughts because of its high population (about 30 million in 1985). The poor spatial distribution and temporal variability of natural resources in the region make small-scale, community-based, low- to medium-level technological approaches ineffective.

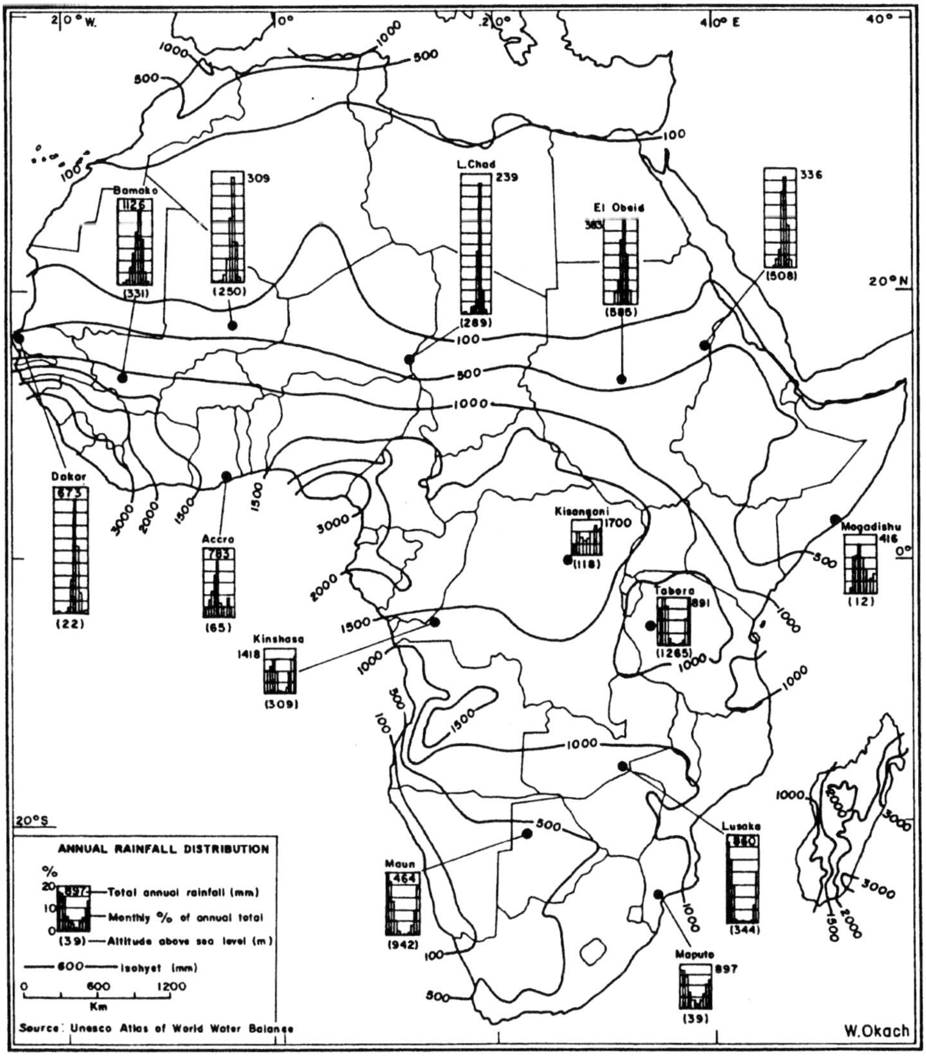

Fig. 1. Annual rainfall distribution in Africa.

Major types of water systems

River systems

The climatic variability has an impact on the runoff characteristics of the continent. Although there is considerable runoff in tropical West and Central Africa, the overall runoff of the continent is lower than the runoff in North and Central America, which have a land area of only 80% of that of Africa. The high rate of evaporation, about 570 mm/year, reduces the effectiveness of rainfall and introduces marked seasonality in the river regime. River systems or basins constitute land area drained by a river and its tributaries and form a framework for understanding the inputs and outputs of the system.

Africa has nine major river basins — Nile, Congo, Zambezi, Okovango, Orange, Volta, Niger, Lake Chad, and Senegal — as well as small ones draining the east coast and discharging their waters into the Indian Ocean. Nearly all are international basins, some traversing more than eight countries.

Total river runoff amounts to 4.2×10^{12} m^3/year, and total stable runoff is about 2.1×10^{12} m^3/year (Endersen and Myhrstad 1987). Africa withdraws about 3% of its annual river flow, but the percentage going to countries north of the Sahara is high. Although the Congo Basin alone has more than 50% of the total river runoff, Africa would require only 0.5–1% of the stable runoff, or a maximum of 1.06×10^{10} m^3/year, to adequately supply its entire population.

The main characteristic of the runoff is its seasonality, which makes harnessing water resources possible only through the use of reservoirs. Most of the reservoirs use capital-intensive means, requiring large amounts of foreign or donor assistance. Moreover, many reservoirs are situated in sparsely populated areas, thus increasing the cost of delivering water to inhabited regions.

Lakes and dams

Lakes and dams regulate the flow of running water. They are also receptacles of sediments and other pollutants. These sediments build up behind the dams and can quickly diminish the storage capacity of reservoirs, interrupt sediment for floodplain fertility, and disrupt the integrity of the delta lands.

The African lakes have a total volume of 30 567 km^3, covering a surface area of 165 581 km^2. Lake Tanganyika alone could supply 400 million people in sub-Saharan Africa through the extraction of only 0.05% of its volume annually. All the lakes in Africa are international, shared by more than two countries, except Lake Tana in Ethiopia.

These lakes contain more aquatic biodiversity than any other lakes in the world. Lake Victoria has more than 300 endemic species; Lake Tanganyika, more than 140; and Lake Malawi, nearly 500 (Master 1990). Pollution, as well as misuse of the water, poses the greatest threat to this aquatic biodiversity.

The continent has 2.4% of the world's large reservoirs (those more than 15 m high). Half of Africa's are in South Africa. These multipurpose reservoirs were mainly designed for hydropower generation, although many are used for water supply, irrigation, industry, and domestic purposes.

However, the development of dams and reservoirs is currently fraught with environmental problems, such as high rates of siltation and evaporation, tropical diseases, weeds, and eutrophication. River water generally carries substantial amounts of suspended matter into lakes and reservoirs. Although these particles could be removed by presedimentation, followed by coagulation, flocculation, sedimentation and filtration, dams and reservoirs require huge capital outlays for construction, operation, and maintenance, sometimes far beyond the means of a single state. Moreover, the mineral constituents in lakes vary considerably, and some lakes, such as Turkana in Kenya, have salinities beyond acceptable limits and require demineralization.

Groundwater systems

Groundwater accounts for more than 95% of the Earth's usable freshwater resources and plays important roles in maintaining soil moisture and streamflow and in sustaining wetlands. Groundwater has several advantages over surface water:

- It has natural protection against evaporation, especially in deep aquifers.
- It does not require storage.
- It can be developed on a small scale for rural and dispersed communities.
- It can be developed at low cost, especially in the case of shallow wells.

However, these benefits are experienced in states that have reliable groundwater recharge. The arid and semi-arid areas of the continent, therefore, are excluded.

Although groundwater is intricately related to surface-water resources through the hydrological cycle, its availability depends on climate, geology, and vegetation cover. Rainfall and geology determine the rate of recharge and are therefore responsible for sustainable supplies. Groundwater, on average, contributes approximately 30% of total runoff, although this proportion varies considerably within different geographical zones. For example, in arid and semi-arid areas, surface runoff is more than 90% of the total runoff.

The main hydrogeological formations (Fig. 2) are sedimentary basins, the crystalline basement-complex rocks, and volcanic formations. The sedimentary basins occur in the arid and semi-arid areas of Africa, the Sahara and the Kalahari. The Sahara basin alone holds more than 60 000 km^3 of water (Burdon 1977) and has high yields and water of good quality. Because recharge is poor, much of the water is fossilized and not sustainable. The groundwater resources appear to be rather more developed in Algeria, Libya, and Egypt, whereas the potentials of the Congo and Kalahari basins have yet to be explored.

The crystalline basement-complex rocks are less productive, but their importance is great because of their geographical distribution. The aquifers occur within the weathered residual overburden, usually about 15–30 m thick, and along fractures that typically extend downward some 60 m (Wright 1985). The rates of recharge vary considerably, although typical rates are 3–10% of the mean annual precipitation (Chilton and Smith-Carington 1984). The richest aquifers produce 12–15% of the total annual precipitation. However, the boreholes drilled in fractures may have very high yields during the wet seasons but generally dry up during long droughts.

The exploitation of regional aquifers, such as the Nubian sandstones covering Libya, Egypt, Sudan, and Chad, requires regional cooperation for standardizing data and information and for improving knowledge of the aquifer.

Water stress

A country whose renewable freshwater availability on an annual basis exceeds about 1 700 m^3/person will suffer only occasional or local water problems. Below this threshold, countries begin to experience periodic or regular stress, a condition in which the annual availability of freshwater is 1 000 – 1 700 m^3/person. When the annual availability of freshwater falls below 1 000 m^3/person, countries experience chronic water scarcity and the lack of water begins to hamper economic development and human health and well-being. When annual renewable freshwater supplies fall below 500 m^3/person, countries experience absolute scarcity.

The annual renewable freshwater available declined from 20 000 m^3/person in 1950 to about 10 000 m^3/person in 1980. Renewable water, unlike nonrenewable water, is continuously renewed within reasonable time spans by the hydrologic cycle. The amount of renewable water available depends on its natural rate of recharge and the rate at which it is withdrawn for human needs.

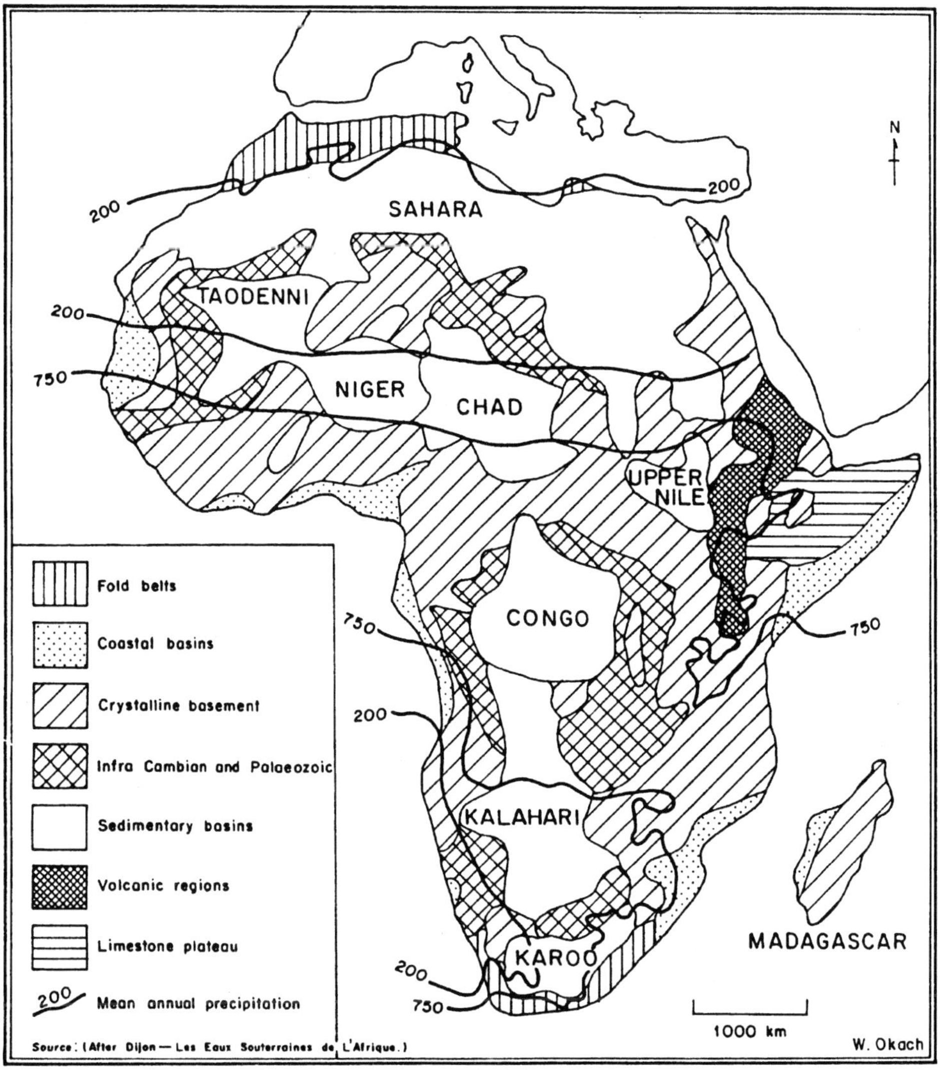

Fig. 2. Main groundwater regions in Africa.

There are abundant water resources available within the humid tropics. Figure 3 indicates countries in which water scarcity or stress will be predominant by 2025. The areas mainly affected — subhumid tropical, arid, and semi-arid countries — are some of the world's most disadvantaged countries (UNSO 1991).

Uses of water

When water supplies were plentiful and most rivers were unregulated, society considered water a free commodity. However, as demand rises, water uses begin to compete for limited supplies, and effective instruments for making rational decisions regarding water-use allocation become important. As demand rises even further, freshwater supplies become a less freely available commodity, as a result of shortage and degradation. As society places more demand on water resources and the cost of producing and delivering water increases, water becomes an economic commodity in socioeconomic development, just like capital and labour. Consequently, over the longer term, investments in water will compete with investments in other forms of production; allocation will be dependent on market forces; and subsidies will be difficult to justify.

There are many competing uses of water, including those for recreation, hydropower generation, navigation, and fisheries, domestic purposes, municipal and industrial supplies, and irrigation. It must be noted, however, that water as a resource must be culturally defined because water by itself is not productive: its use requires some minimum level of social infrastructure for it to be productive. Indeed, lack of such infrastructure is the main hindrance to water-resource management and effective water use in developing countries.

Municipal and rural domestic water supply

Although the United Nations (UN) International Drinking Water Supply and Sanitation Decade (1980–90) set out to supply 500 000 people with safe drinking water and sanitation, by 1993 more than half of the African population had neither safe drinking water nor sanitation. The variations from country to country were staggering. For example, in the urban water-supply sector, almost 80% of the urban households in Egypt received piped water, but only 7.2% received this in Burkina Faso. Water supply from public stand posts ranged from 2.3% in Madagascar to 59% in Botswana. Rural water supply was predominantly from community stand posts throughout Africa. The daily level of water-supply services also varied, from 15 L/person in Angola to about 270 L/person in Madagascar; in the rural areas, the average varied from 20 to 40 L/person.

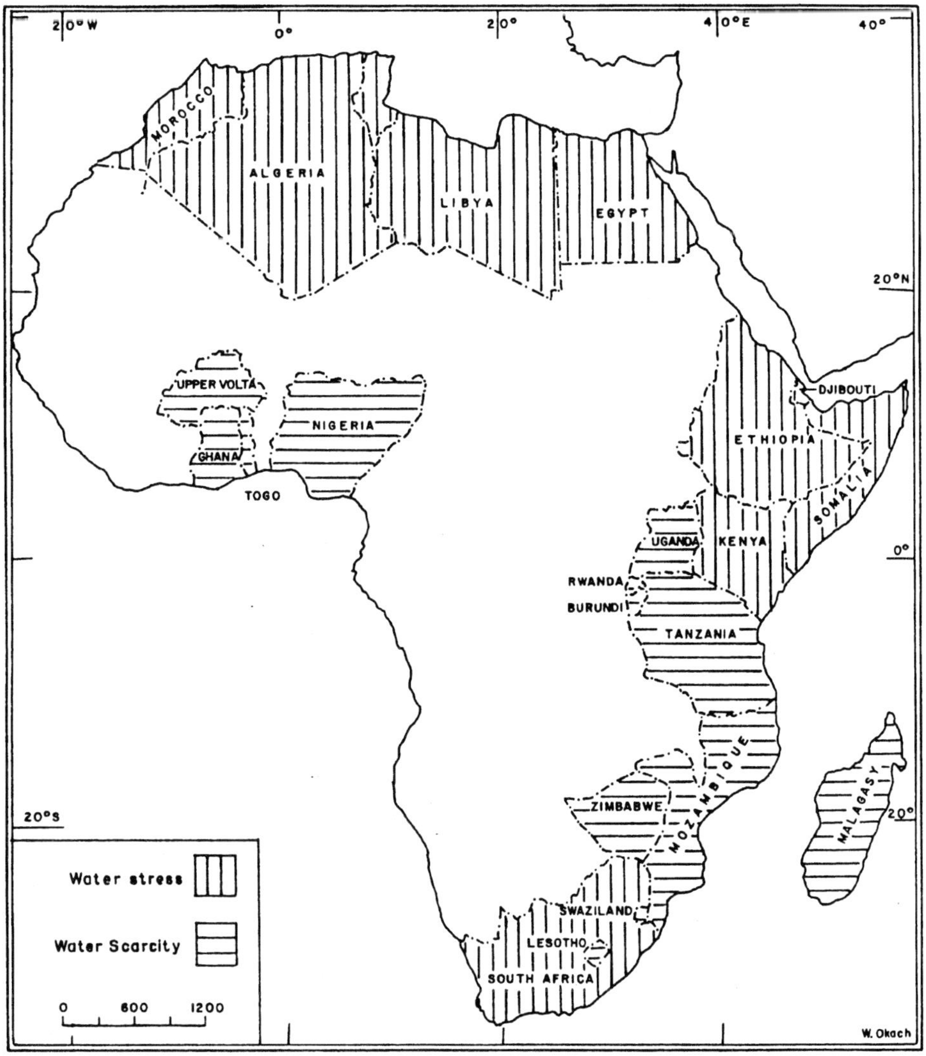

Fig. 3. African countries projected to experience water stress or scarcity by 2025.

The population of cities in Africa is growing by 6–9% per year, and the demand for increasing amounts of water can no longer be met by sources within the urban areas. Thus, unavailability of water, as well as heavy pollution of water sources in and around cities, has made basin transfers from distant locations inevitable. In this respect, the cost of producing and delivering a unit of water to urban areas has not only escalated but has also taken water away from other competing users. Although many countries are expanding their urban water systems, rural areas continue to suffer from neglect. The conflict related to reallocating water from rural to urban areas is growing, but no mechanisms for solving such problems presently exist.

Agriculture

Africa's agricultural potential is vast. Currently, only 24%, or 2.817×10^9 ha, of the arable land is under cultivation. About 68% of Africa's economically active population is currently engaged in agriculture but contributes only 24% of the gross domestic product (GDP), although regional variations occur (World Resources Institute 1993).

Irrigation takes up 88% of the total water withdrawals but represents only 1.9% of the cultivated land in Africa (UNECA 1989). With the greatest part of African agriculture under rainfed crop farming, food insecurity is largely caused by variability of rainfall. Moreover, about one third of the continent has a mean annual rainfall of less than 700 mm, which is too little to support rainfed crop production.

From 1975 to 1979, Africa produced 83% of her cereal requirements and imported 8×10^6 t. However, it is estimated that by 2000, the net import will have risen to 49×10^6 t, with only 56% produced locally. Because 53% of all irrigated land is under cereals, the importance of irrigation to obtain food security must not be underestimated.

The current emphasis on irrigation in Botswana, Burkina Faso, Ghana, Kenya, and Somalia is aimed at rehabilitation and expansion of old schemes that have not made a sustained contribution to food security because of improper management. In Egypt, some efforts are being directed toward improving irrigation efficiency. Moreover, African countries involved in the Interstate Committee for Drought Control in the Sahel plan to bring nearly 1.0×10^6 ha under irrigation by 1998. The construction of Nanatali Dam on Senegal River will increase Senegal's irrigable land by 400 000 ha, and Mali's by 300 000 ha. Egypt and Sudan will irrigate a further 2.0×10^6 ha in the Nile Valley, and 400 000 ha will be irrigated in Uganda, Tanzania, and Kenya.

In the dry areas of Africa, irrigation consumes 100 m^3 of water per tonne of biomass produced (Falkenmark 1992). Increasing biomass production will deplete the amount of water available for other uses. Moreover, irrigation water is frequently used for growing rice and maize or for grazing pasture, which, compared with cotton or horticultural crops, have relatively low values, particularly because national food policies tend to subsidize the cultivation of such low-value crops. Only Botswana has considered the cost of irrigation water realistically and has decided that it cannot pursue the policy of self-sufficiency in food security.

Industry and mining

Industries require large amounts of water. The average consumption for industry in Africa is 5% of the water withdrawal, but the actual amount depends on the extent of the manufacturing and the type of industry. Textile and steel mills, canneries, tanneries, and breweries, which are predominant industries in Africa, are some of the heaviest water users. Except in food processing, about 80% of water used by industries is for cooling and cleansing and, therefore, often does not need to be of high quality.

More than 30% of the world's mineral resources are in Africa. Open-cast methods of mineral extraction are prevalent. The expansion of mining activities in the Pretoria–Witwatersrand–Vereeniging region in South Africa, where 60% of industrial production is concentrated, has led to the development of transboundary water resources from rivers emerging from the Lesotho Highlands and the Komati River in Swaziland (Forster 1993). Further expansion of the water supply is planned for Tswasa and the Limpopo.

Sources of stress and stress indicators

Water is important in socioeconomic development. The presence or absence of water stress is, therefore, indicative of whether a nation has adequately or inadequately managed its resources, and failure manifests itself through acute water shortages and rising conflicts among the various water users.

Water problems may be grouped into two broad categories: those caused by contamination and those caused by overexploitation.

Contamination, especially of groundwater, is usually difficult to detect, and monitoring is costly, time consuming, and not always effective. The contamination is usually not detected until noxious substances appear in drinking-water supplies, as in the Bamako water supply in 1993 (Heidenreich 1993), at which point pollution has already dispersed over a large area. Evidence shows that freshwater quality is declining everywhere, with contaminants coming from agricultural, urban, and industrial land uses. The growing contribution of non-point sources of contaminants, such as fertilizers, pesticides, septic systems, street drainage, and air and surface-water pollution, has challenged known cleanup methods developed for large point sources.

Overexploitation of freshwater resources occurs in areas where there is scarcity. People are increasingly moving into areas of marginal productivity. Ideally, sustainable groundwater extraction must not be greater than recharge, if recharge occurs. However, in arid and semi-arid regions of Africa, more than 40% of the total groundwater is fossilized water. When average groundwater withdrawals exceed the average recharge rates for an extended period, aquifers become depleted, and the water table or water pressure begins to drop, which leads to the following:

- Shallow wells dry up.
- Production wells must be drilled to progressively greater depths, requiring more energy for pumping.
- Aquifers in coastal areas can become contaminated by saltwater intrusion.
- Subsurface materials may gradually compact and cause surface subsidence.

Diagnostic surveillance for risk assessment is only vaguely defined in Africa. There are no inventories of the sources and magnitude of pollutants or monitoring of pollutants. Simple technologies are currently used, although such methods require surveillance at the source; other methods include the use of a very narrow range of chemicals for domestic water supplies. Some are useful at the tap or source of distribution. The best surveillance methods to adopt will depend on the size of the basin and the type and extent of the pollution.

Long-term risks resulting from these water problems include

- water shortages, requiring expensive interim measures like developing new water sources or adapting sophisticated treatment measures;
- health hazards, such as exposure to pathogens, carcinogens, and nitrates;
- damaged ecosystems;
- structural damage and coastal flooding, as a result of land subsistence;

- economic and social hardships, especially affecting the rural poor; and
- political conflicts, diverting resources from productive investments.

Population growth

Africa's population has more than doubled in three decades, from 220.3 million in 1950 to 470 million in 1980. Between 1981 and 1990, 110 million people were provided with safe drinking water, but the population increased by 140 million, thus diminishing progress during the UN International Drinking Water Supply and Sanitation Decade. Presently, 54% of the continent's population does not have safe drinking water, and 64% has inadequate sanitation. Moreover, there is evidence to show that 80% of endemic disability is caused by waterborne diseases and that water scarcity and low incomes are related in many parts of Africa.

The effects of high rates of population growth on water-resource management are twofold. On the one hand, the cost of producing and delivering a unit quantity of water will be enormous just to keep the present levels of service, thus removing investment opportunities from other sectors of development. On the other hand, development along traditional lines will become ever more difficult as demand pressure on water grows, leaving less water on a per capita basis. Although there is no one-to-one relationship between population growth and higher water requirements, it is evident that with substantial increase in population, total water requirements for various uses will increase.

Human settlement

Urbanization greatly influences water quantity and quality because of runoff and sewerage. Land-use changes, such as impermeable surfaces replacing permeable ones, increase the volume, velocity, and temperature of urban runoff; reduce the base flow of rivers during dry periods; raise the temperature of urban streams; and concentrate pollutants. Untreated sewage, on the other hand, introduces large quantities of nutrients, pathogens, heavy metals, and synthetic organic chemicals into surface water.

In most African cities, sewage treatment is poor because the plants are old, technical knowledge is limited, operation and maintenance costs have escalated, and water standards are poor or nonexistent. About 95% of urban sewage is discharged into surface water without treatment. With about 2×10^6 t of unguarded and unmanaged solid waste generated every year (WRI 1993), bacteria, parasites, and viruses in water supplies remain a more serious threat than toxic contaminants. The problem of waste management is further aggravated by the development of unplanned slum settlements, some of which house about 50% of the urban population, as is the case in Nairobi.

A major sewage-treatment problem is the lack of separation of various wastes (household, industrial, and hazardous) that would facilitate recycling or reuse. In addition, there are thousands of poorly maintained or simply uncontrolled dumping sites, contributing significantly to contamination of local aquifers, lakes, and rivers. In Maputo, Mozambique, for example, nitrate contamination is far beyond World Health Organization (WHO) standards. The local authorities have neither sufficient funds, transport, and technical expertise nor the water-management infrastructures to deal with such solid waste in urban areas.

Industry and mining

Industry employs about 12% of the active population in Africa and contributes 36% of the GDP. Many African industries involve wet processing of agricultural commodities. Industrial pollutants include dust from smelting and metal processing; heavy-metal solutions used in plating, galvanizing, and pickling; metals and metal compounds used in paints, plastics, batteries, and tanning; and leachate from solid-waste dumps. Heavy metals and synthetic organic chemicals accumulate at higher levels of the food chain, thereby posing a special risk to humans and aquatic systems. Such pollutants increase the risk of cancer and reproductive anomalies in humans, fish, and aquatic mammals.

Although industrial sources of pollution in Africa are negligible compared with those in developed countries, the absence of standards and lack of monitoring make any form of industrial pollution significant. In many countries, highly toxic industrial waste and chemical-liquid wastes are discharged into municipal sewerage systems. Pretreatment of industrial effluent is generally nonexistent or insufficient for industries that use old technologies. Breweries, textile mills, and tanneries are everywhere, and they discharge wastes of relatively high loading, for which the organic removal efficiency decreases rapidly (Otieno and Kilani 1991). Similarly, mining and petroleum extraction pollute freshwater, either through discharge of brine or through leaching from mine tailings into the groundwater, and their impact is likely greater than that of manufacturing or processing.

Agriculture

Agriculture causes water stress both through use of water and through runoff of agricultural inputs. Because of the need to feed more people, forests and grasslands are converted into farmlands, often increasing soil erosion, sedimentation, and the use of fertilizers and pesticides. Frequent ploughing and excessive irrigation also pollute freshwater systems with sediments, salts, and pesticides.

Agriculture is responsible for 24% of the soil degradation in Africa; overgrazing, common among the pastoralists in the arid and semi-arid areas, contributes 49% (Ayoub 1993). Livestock grazing removes vegetation, compacts the soil, and generates large quantities of manure, although high-potential areas with abundant rainfall generally do not suffer much from overgrazing.

Diversion of irrigation water causes deterioration of water quality. Irrigation consumes a large proportion of the total water supply, although the volumes consumed vary slightly from country to country. Egypt, for example, uses 84% of its supply on irrigation for 95% of the country's agricultural production. However, evaporation accounts for an additional 4% of water use, which generally is not taken into consideration. Such large water diversions create scarcity, distort the market for water, and raise issues of equity. Because the cost of water delivery is shared by all consumers, the consequence is that small consumers subsidize large ones.

In irrigated areas, the return flow and drained water can pollute freshwater by introducing fertilizers and pesticides that cause eutrophication and algal growth. Besides destroying habitats and reducing biological diversity through unfair competition of species, the pathogens transmit diseases to humans. In addition, excessive irrigation contributes to salinity and raises the water table, although the higher water table may be used for recharging groundwater, which in turn may improve or encourage conjunctive water use. In South Africa, the lower portions of major rivers, such as Modder, Rietz, Vaal at Douglas, and Orange, are at various stages of salinization as a result of irrigation and urban development (Herold and Kakebeeke 1991). No irrigation scheme in arid and semi-arid areas can be successful for long unless it has adequate drainage.

Land-use changes

Deforestation

Cutting trees mainly for fuel, building materials, and farmland and urban extensions is responsible for about 14% of soil degradation in Africa (Ayoub 1993). In West Africa, forests are disappearing at a rate of 2.1% per year, and in Central Africa, deforestation occurs at a rate of 0.6% per year, mainly as a result of the need to clear land for grazing. This deforestation leads to stream-bank erosion, stream sedimentation, and loss of habitat for a wide variety of organisms. In East Africa, increased sunlight, combined with nutrient enrichment in streams, is promoting growth of the filamentous algae favoured by schistosomiasis-carrying snails and of weeds that are threatening freshwater lakes. Deforestation also changes the hydrological cycle, especially the rate of rainfall infiltration and evaporation, soil moisture, and temperature conditions, although the magnitude of each of these changes depends on the nature of the new land use. Deforestation, therefore, has the potential to contribute to climate change.

Property rights

Ambiguous or nonexistent property rights constitute one of the main causes of land degradation in Africa. The semi-arid and arid areas are inhabited by pastoral groups that, over the years, have been grossly marginalized. The phenomenon is widespread and requires immediate attention so that desertification can be arrested and the impact of drought mitigated. Research should be focused on integrating land–water relationships, analyzing human impact on the hydrological cycle, and improving the efficiency of water uses.

Land-use planning

Land-use planning cannot be separated from the issues of water supply and the health of freshwater systems. Six phenomena provide the linkages between land and water:

- rainwater partitioning in the hydrological cycle;
- water's mobility and capacity as a solvent (which, on the one hand, guarantees the effective transport of substances needed for life and stimulates both biotic and abiotic diversity and, on the other hand, spreads pollution and causes the deterioration of water quality);
- societal use (which always results in the return of water used for industry, agriculture, and households to natural systems, carrying with it a load of pollutants);
- water's erosive capacity;
- soil fertility; and
- pore water in soil (which links soil drainage and groundwater decrease with the soil cover and land surface).

Currently, these links between water resources and socioeconomic development are at best vaguely understood. In most cases, water development, rather than management, is emphasized; environmental degradation is not integrated into water-resource planning; integrated riverbasin organizations are weak and highly politicized; and gaps between water users and water ministries are wide.

Drought and desertification

Drought, more than any other single factor, has caused food insecurity and water shortage. Major droughts occurred in the 19th century; the most dramatic one, in the 1980s, affected 20 countries and some 30 million people in sub-Saharan Africa. In Gambia, for example, drought reduced recharge of the groundwater, necessitating further deepening of wells. These wells then collapsed, aggravating the competition for water. Subsequent overuse of wells caused salt-water intrusion in those boreholes that survived the drought period.

Although research continues to decipher the climatic trend, people in high-risk areas must evolve a drought and disaster management strategy, establishing early-warning systems, strengthening institutions, and removing obstacles to the free movement of food. Early-warning systems have already been established in 30 countries, mainly along the Sudano-Sahelian zone and in East and Central Africa, with the assistance of the Food and Agriculture Organization of the United Nations and of other bilateral aid agencies.

Summary

Natural occurrences (such as drought, desertification, and climate change) and human activities (such as agriculture, population growth, industrial development, and land-use changes) are causing untold pressure on freshwater resources. These pressures have caused both environmental deterioration (including pollution of freshwater systems) and overexploitation or deterioration of important water catchments, resulting in lowered groundwater levels. The magnitude and major cause of stress vary from one region to another. In North Africa, where the levels of irrigation and industrialization are high, water shortages appear to result from scarcity and overuse. In the Sahelian region and East and Central Africa, water scarcity appears to be due to drought. In southern Africa, water shortage is due to drought and overuse. Because many countries in northern and southern Africa have mineral resources and are fairly industrialized, their approaches to tackling water shortage will be different from those of the rest of the continent.

Water stress has several repercussions. Socially, human health is at risk, and water-related conflicts are imminent. Economically, the cost of producing and delivery of a unit quantity of water has escalated, thus diverting investments from other productive areas. Environmentally, sustainability of the ecological systems is threatened by overexploitation or pollution.

Some water-related problems are local, but others are regional. Most of the water systems in Africa that are able to endure marked seasonality in climate are international, whereas local water sources that are generally prone to drought are local. Because nearly two thirds of the continent is arid to semi-arid, poverty and insufficient agricultural production put pressure on freshwater management almost across the continent.

Each type of risk requires different assessment methods and solutions. A large proportion of Africa's population is affected by domestic-water shortages. Although the source of drinking water is not always the same as that of industrial or irrigation water, there is a growing tendency to divert water from domestic use to industry or agriculture.

Responses to stress

International responses

Research and training

International organizations have played a major part in catalyzing development in the water-resource sector, creating awareness and focusing sharply on the problems and challenges in that sector by contributing funds; providing technical assistance and training; and facilitating research, networking, and information dissemination.

Enhanced awareness of the problems, challenges, and opportunities in the sector in Africa has been accomplished to some extent through international conferences, from which workshop proceedings, protocols, and binding statements have been produced. From these outputs, the UN Economic Commission for Africa and other UN agencies have assisted countries with follow-up. The Lagos Plan of Action provides one example of increased emphasis on water-sector issues.

Current freshwater research focuses on two key issues:

- the need to integrate population growth with sustainable water use (high rates of population growth are diminishing the impact of any successes that have been achieved); and
- the need to integrate land use with freshwater systems, thus liberating the water sector from structures that inhibit participation by resource users and the private sector.

Further elaboration of the catchment approach to water-resource planning is needed.

Current irrigation research focuses on system performance, water management, irrigation organization, institutional change and policies, farmers and the farming community, and the environment. The World Bank, the UN Development Programme (UNDP), and the International Commission on Irrigation and Drainage, have funded the establishment of the International Program for

Technology Research on Irrigation and Drainage (IPTRID). The mandate of IPTRID is to coordinate and focus research activities on the efficiency of water use in irrigation, the modernization of irrigation and drainage systems, and the improvement of maintenance technology. Also needed is research in biotechnology to isolate high-yielding crop varieties that consume less water.

National responses

Reallocation of water from irrigation to industry

Because irrigation consumes more than 80% of all the water, there may be an argument to be made for the reallocation of water to urban households and industry. However, government commitment to food security or, in some cases, lobbying by certain interest groups may make reallocation difficult. In general, African urban water supplies are better developed than rural ones, partly because urban areas are able to raise funds and reduce marginal costs. Urban areas also have adequate institutional structures, through local authorities.

One way to make the reallocation of water workable is to set the tariffs paid by agricultural users high enough to force them to conserve.

Population policy

Population mobility and rural–urban dynamics are affected by national policies on rural development, land tenure, and equitable sharing of resources. Rural development is important because more than 80% of Africa's population lives in rural areas that are poor and lack infrastructure and employment opportunities. The issue of land tenure is problematic in much of sub-Saharan Africa. The poorest, marginalized farmers have often been forced to sell fields to bolster their incomes and have subsequently had no choice but to overfarm their remaining land, thereby contributing to ecological degradation. Newly landless farmers, on the other hand, migrate to the urban centres and swell the ranks of the unemployed there. Security of tenure is fundamental to achieving rational use of water in the wider context.

There is a need for a national population policy that takes into account available freshwater resources and the rate at which such resources may be developed. In addition, priorities must be defined to target specific populations for increased water activity. At present, it seems expedient to target poor rural people, small-scale entrepreneurs, and minorities.

Methods to improve efficiency and reduce water losses

Different methods have been adopted to improve efficiency. For example, in Conakry, Guinea, 50% of the water pumped in 1988 was unaccounted for, and only 10% of the bills were collected. Today, only 25% of the water is lost, and 85% of bills are paid. In Lomé, Togo, only 20% of the water is now unaccounted for, almost all bills are collected, and water subsidies are not required. This efficient management is attributed primarily to the commitment and job satisfaction of the employees. Although water management in Conakry is undertaken by a private company, whereas that in Lomé is undertaken by a public utility, in both cases there is suitable environment for efficiency, as the companies have streamlined their operations.

Water systems have tended to operate more efficiently if the responsibility for daily operations and that for capital investments are kept separate. Such is the case in Guinea, where 51% of the company is privately owned and the remaining 49% of the stock is held by a parastatal, which supervises investments. By keeping the two functions separate, there is less likelihood of corruption. The parastatal is autonomous and, therefore, does not depend on subsidies. However, there is no guarantee that these measures will always bring the benefit of efficient management.

Water-resource planning

Planning requires adequate information for understanding both the hydrologic systems on which water-resource management is built and the nature of the interactions between the natural and the socioeconomic systems. Collection of data on the quantity and quality of water resources, on the demands and supply, and on conservation measures is essential for proper management. African countries were drawn into hydrological networks in the 1960s and 1970s, through international cooperation, and the information gathered improved the rational use of water resources. However, current challenges associated with the dynamics of socioeconomic development, including budget deficits, reduced government spending, inflation, and increased price for raw materials, have caused deterioration of these hydrometeorlogical networks. Programs for monitoring sediment discharge, pollution control, and water usage have become inadequate or are in disuse. Currently, only 20 African countries have some kind of integrated environmental-management plan. In many cases, such plans are used as shopping lists for donor investments, and there is little actual implementation.

Community responses

Involvement of consumers

Participatory approaches to water-resource management require consumers to play a part in the decision-making process. Consumers who know the status of the resource and are informed about the limitations of their actions will make rational decisions. Emphasis on natural-resource management is thus being devolved to local communities and authorities. The process is slow, but it is expected to become more popular.

Water-management issues

Water policy

African nations are aware that water scarcity is a problem. Most countries have a ministry of water affairs and a parastatal dealing with commercial water supply. In addition, there are usually several institutions dealing with water supply and related issues, such as waste management and health. However, the mandates of these diverse institutions tend to be fragmented and unclear, and linkages among them are weak. Only Morocco has a central water board with a mandate to formulate policy.

A national water policy can chart a clear direction and provide the necessary linkages between the water sector and critical areas of the economy, including health, agriculture, sanitation, and rural development. This would eliminate conflicting priorities that otherwise may exert unnecessary pressure on the resource.

There are several levels at which a national water policy can be enacted. At the highest level is the overall water policy, such as exists, for example, in Botswana and Morocco. The necessary instruments to support the policy are those that define water use; institutional responsibilities and coordination; issuance of permits for water abstraction and use; treatment of effluent and its safe discharge; groundwater exploration; water abstraction and use; tariffs; population and water use; water and health; the role of women and popular participation; wetland and aquatic ecosystems; international water courses and regional cooperation; and training and registration of water scientists.

In general, national water policies are almost nonexistent in Africa, but where they do exist, they are weak, poorly stated, or unimplemented. In other cases, personnel may be unmotivated and poorly trained.

Legislation

All African countries have laws that require permits or licences for abstracting water from either surface or groundwater sources and for discharging waste or effluent into a water body. However, these laws are often inadequate for the following reasons:

- Legislation governing rational use and management of water resources is scattered in various enactments, and responsibilities for implementation and monitoring are also scattered in various government departments.

- Complementing institutions are poorly organized and have no resources for monitoring, policing, and punishment of offenders; field and laboratory equipment and logistical support are lacking; and human resources are unmotivated.
- Rules and regulations have not been elaborated to provide a firm basis for applying the laws.
- Appropriate quality standards have not been set for effluent discharges.
- Penalties for offenders are lenient and constitute an inadequate deterrent.

Botswana is one country that revised its entire water law in 1990. Its experience suggests that satisfactory legislation depends on the prior definition of a water policy, and the success or failure of the legislation must be measured by the extent to which it succeeds in helping to fulfill the objectives of the water policy. A framework of integrated land and water management is needed to ensure harmonious allocation and development of water and land resources, both for increased supply and for agriculture, forestry, recreation, and urban and industrial development. In turn, integrated land and water planning based on an agreed water policy with a sustained long-term strategy should be integrated with overall socioeconomic planning, taking into account population growth and national needs and aspirations. Such policies and legislation require public consensus and political commitment.

Development of human resources

Training institutions in Africa and elsewhere have produced many water scientists and administrators, but the shortage of skilled human resources remains serious. Many trained Africans seek better opportunities elsewhere. A 1988 UN report indicated that there were about 70 000 African professionals working outside the continent; at the same time, about 80 000 expatriates worked in Africa.

Most countries have gone through educational reforms since independence. To meet human-resource needs of the new nation-states, educational institutions were expanded in scope and coverage. In addition, there has been constant exchange of technical training and expertise, not only among African countries but also between these and developed countries. However, since the early 1980s, the level of spending on education has been going down in many countries.

Human-resource-development policies need to be strengthened to ensure more effective use of trained personnel and improved employment conditions. Many countries in the region are reforming their civil service to prune the workforce and reward productive workers. However, the results of these reforms will take time.

Demand management

Demand management encompasses mechanisms by which water is allocated efficiently to different user groups. This requires that the rules and regulations governing allocations be known. Cost recovery and tariff policies should take into account expenditure investment, operation and maintenance, system expansion and replacement, and return on investment. In countries where the state dominates the water sector, policies must also take into account potentially adverse effects on the poor and disadvantaged sectors of the population.

This concept is particularly relevant in societies with scarce water resources, where choices about allocation may preclude opportunities for some people. This concept may also be used to increase awareness of the need to harness scarce and vulnerable water resources. It provides a means of attaching value to water and estimating how alternatives may facilitate or hamper the achievement of desired objectives.

Price mechanisms

Although water has traditionally been considered a common good, shortages have occurred and the cost of getting water from alternative sources has increased, so its value has gone up. At the same time, there has been reduced investment in water infrastructure in developing countries. However, external funding for water projects has increased or remained constant over the past decades, thereby giving a false value to water as a resource. Disbursements from the UN increased from 31 million United States dollars (USD) in 1973 to 184 million USD in 1985, representing an annual growth of 16.9%. The highest rate of growth was in Africa, Asia, and the Pacific, where 50% of disbursements went to drinking-water supply and sanitation and 24% went to irrigation. Loans and credits for water projects from the World Bank – International Development Agency and the International Fund for Agricultural Development increased from 504 million USD in 1976 to 1 748 million USD in 1985. Of the total disbursements of almost 11 billion USD over the period 1976–85, 54.6% went to agriculture; 28.7%, to drinking-water supply and sanitation; 15.4%, to hydropower; and 1.3%, to navigation. Loans for irrigation were mainly made to the countries north of the Sahara with proven irrigation cultures, whereas loan disbursements for water supply were mainly for urban water, thereby explaining the disparity in service between urban and rural areas.

In the early 1980s, an analysis of the tariff policies of 31 African countries showed that 28 had one-tariff policies for urban and rural areas. For the urban areas, 18% of the countries had policies aimed at full-cost recovery; 75% had tariffs partially covering cost; and 7% did not impose any tariff at all. Absence of cost-recovery measures meant that 82% of the countries' urban water systems relied on the central government to subsidize their operation and maintenance costs. For the rural areas, all countries charged some tariff, but 46% charged tariffs that covered only part of the operation and maintenance costs. The implications were as follows:

- Almost all rural water systems depended on central-government revenue to meet their operation and maintenance costs.
- Even where such subsidies existed, they were inadequate, and there was a need for upward revision to cover mounting costs of producing and delivering water.
- It was not clearly defined what costs the tariffs covered — investments, operation and maintenance, system expansion, plant renewals, or various combinations of these.
- The subsidies were established without targeting the disadvantaged segments of society.

There are many opposing views about pricing water. Politicians often argue that water must be cheap to ensure that the poor have access to it. Frequently, the poor do not benefit from low tariffs, partly because they lack water connections and they usually buy water from vendors at prices 10 times more per litre than those paid by people with household connections. High-income groups tend to benefit from low tariffs. Moreover, when water charges are low, people tend to use it carelessly. The poorest segment of the population consumes 15–20 L/person per day, compared with the 50–125 L/person per day consumed by the wealthier people with direct connections to water systems. Under these conditions, the poor inadvertently subsidize the wealthy because funds needed to cover water-company deficits are shifted from other social programs, such as education and health, which might otherwise benefit the poor.

The cost of production and delivery of a unit quantity of water includes initial investment, operation, and maintenance costs. The price of water or tariff charged to consumers should include costs plus interest on capital, depreciation, expansion, and return on assets. Using this framework, one can cost each new system that is built and then establish appropriate tariffs. Alternatively, costs incurred in setting up new systems can be added to costs of existing ones, for a method of average costing and pricing. Either method has advantages and disadvantages, depending on the political situation, rural–urban balance, and the established institutions that provide water and sanitation services.

However, the costing and pricing of water do not take into account the willingness to pay. The willingness to pay depends on the beneficiaries' perception but does not necessarily go with the ability to pay, which depends on income. The World Bank estimates that the cost of water should not exceed 5–6% of the incomes of the poorest households. This puts a ceiling on whom may be charged a particular tariff. Recently, as a result of inflation and depreciation of local currencies, the cost of production and delivery of water has increased considerably, thus aggravating the problem of cost recovery.

The problem of cost recovery in water supply must be approached from an integrated planning framework that combines three considerations:

- the need to supply basic needs at affordable costs, sufficient for the maintenance of public health and appropriate social dignity;
- the need to recognize the market imperfections resulting from the behaviour of consumers who are either insulated from price mechanisms or not provided with public education on the cost implications of certain water-use habits; and
- savings that can be achieved from improved water control in the reticulation system (accepting some reduction in reliability of supply at times of water shortage).

However, there are no data to quantify price elasticities of water demand.

Options for improving water supply

Several attempts have been made to respond to increasing demand for water. In the past, the tendency has been to increase supply in line with demand. However, caution has to be exercised to limit the amount of water being extracted. The repercussions of responding to demand elasticity, or continuing to provide water in proportion to population growth and other factors without restraint, could have irreversible effects. Whatever limited success could be achieved in arresting apparent problems would also cause severe economic and environmental damage.

Reduction of environmental effects

The notion that prevention is better than cure is generally considered the least expensive and most advisable approach to maintaining water quality. Traditional end-of-pipe waste-management strategies, based on cleaning up water sources and fining offenders, can be combined with tighter government regulations and economic incentives. Two options open to polluters include reducing the volume and toxicity of wastes by recycling; and redesigning processes and products. Institutions that undertake waste-reduction programs often save money by using materials and energy more efficiently or by reducing the costs of conventional pollution control and waste disposal. Transnational companies that operate in Africa can often obtain appropriate technologies for waste reduction from their headquarters in the industrialized countries. However, lack of appropriate legal and economic instruments have hindered progress in this area.

Clear policy instruments are also needed to control runoff pollution. Soil-conserving agricultural and forestry practices, road management, land-surface roughening, and redesigned streets in urban centres are some land-management techniques that can reduce pollution in runoff areas. Improved agricultural management can considerably reduce runoff containing pesticides, fertilizers, and sediments. Other harmful forms of runoff can come from livestock farms and logging areas. Enhanced management activities might include the creation of buffer zones along the riverbanks, dams, and lakes, the use of pretreatment ponds, and the introduction of the biological control of pests. These could reduce the agrochemicals discharged into the water systems.

Various economic incentives or disincentives can also be aimed at waste reduction. These include increasing the taxes on pesticides and fertilizers, increasing the fees for irrigation water, and removing the production subsidies. Production-tax rebates to encourage clean technologies and waste processing, subsidies to encourage environmentally friendly activities, and user charges for municipal-waste collection to encourage waste separation and recycling are some of the many economic incentives that countries may adopt. Unfortunately, poor communication among scientists, planners, and politicians is responsible for the lack of understanding of the potential benefits of such economic instruments.

Pollution-permit trading has been introduced in the United States as a means of reducing pollution. Under this system, the government issues a fixed number of pollution permits, which are then bought and sold at market prices by firms. This can be feasibly applied to tanneries, abattoirs, canneries, and textile mills, some of which are the substantial polluters. However, fixed levels of permissible pollution must first be established by governments. In many countries, such pollution standards are nonexistent, monitoring mechanisms are very poor, and pollution regulations are impossible to enforce.

The goal of all these pollution-reduction activities is to improve the quality of decision-making in integrated water-resource management. It will be possible to take full advantage of the water-reuse option only when it is economically, socially, and politically feasible.

Use of marginal water and reuse of wastewater

Even when the most economical state-of-the-art methods are used for cleaning wastewater, only 80–95% of harmful materials are removed. This still leaves 5–20% of the stable pollutants in the water. However, different uses or reuses may be made of low-quality water. Simple technologies exist to reclaim wastewater for reuse. For example, treating sewage in wetland and fish ponds offers tremendous potential for constructive wetland use. The only requirements in addition to those of traditional wastewater treatment are land and alternative uses for reclaimed water. In Algeria and Tunisia, drainage water from irrigated fields is already being recovered and reused in irrigation systems, and in Egypt, Libya, Tunisia, and Morocco, UNDP has pilot projects to study the technical, economic, and social feasibility of using treated wastewater.

Within Africa's growing cities, great quantities of water will be used and discharged into the environment. Greater Cairo generated 0.9×10^9 m^3 of wastewater in 1990, for example; this is expected to increase to 1.93×10^9 m^3 by 2010. Similarly, Morocco and Tunisia will discharge 555×10^6 and 227×10^6 m^3 of wastewater, respectively, by 2000, and all the urban centres in Botswana combined will produce 66.4×10^6 m^3 of wastewater by 2020 (CEP 1993).

The level of wastewater recycling in Africa is very low: Zimbabwe and Namibia, for example, recycle only 10–25% of their effluent (CEP 1993). However, wastewater is starting to be used for irrigation and aquaculture. In Tunisia, about 3 000 ha has been irrigated, under controlled conditions, with secondary-treatment wastewater effluent; this uses about 7×10^6 m^3 of treated effluent per year. Tunisia's stated policy (like Botswana's) is to achieve 100% reuse, and its ultimate goal for 2010 is to irrigate 30 000 ha with treated wastewater. Wastewater is also used to irrigate the greenbelt around Khartoum, Sudan, and 1 000 ha of Egyptian farmland.

UNDP is supporting research on wastewater reuse to develop treatment strategies that improve the cost effectiveness of reuse and provide systematic quantitative evaluation of the costs of water supply, treatment, and disposal.

Wastewater reuse, especially around major cities, has three advantages:

- creation of a new water resource at the local level that could be used without significant negative impact on the environment;
- decrease in the cost of treatment and disposal of wastewater; and

- generation of additional revenue through the sale of treated wastewater and sludge to farmers to cover the cost of treatment and disposal.

Morocco and Tunisia, for example, charge farmers fees for using reclaimed wastewater (Khouri 1989).

Desalinization of seawater

Desalinization of seawater, a capital- and energy-intensive source of freshwater, is growing in importance in oil-rich countries. The number of desalinization plants worldwide increased throughout the 1980s and early 1990s. Production capacity since 1970 has increased roughly 13-fold, from 13.3×10^6 m^3/d to 18.7×10^6 m^3/d (WRI 1992–93). The distilling process and the reverse-osmosis method are used. Desalinization is still three to four times more expensive than conventional methods of obtaining freshwater. Some of the plants are also being used for treating effluent water or river water to obtain water for boilers. Both groundwater polluted by nitrates and pesticides and municipal water are treated to make ultrapure water for the electronics industry. Egypt, Libya, Tunisia, and Morocco have desalinization plants.

Regional cooperation

Effective water management requires a broad plan for an entire riverbasin. Most river basins in Africa are shared by two or more countries. Cooperation in water-resource management needs to be pursued at two levels: (1) through comprehensive management of domestic supply and demand; and (2) through comprehensive regional planning and arrangements to import water from surplus countries to deficit countries. Competition and conflicts in water use generally emerge when there is lack of cooperation, leading to international tension. International law on shared freshwater resources is limited to ensuring that the activities of upstream nations do not conflict with those of downstream nations. The role of the Organization of African Unity in arbitrating water disputes is not clear. As a result, many downstream countries pursue their rights through diplomacy.

Regional treaties and protocols concerning water resources have been negotiated. There are 54 transboundary river–lake basins. Among these basins, only a handful — Senegal, Gambia, Niger, Chad, and Kagera — are overseen by some form of intergovernmental organization charged with the exclusive task of planning for integrated development of natural resources, energy, and other water-related infrastructures. But these organizations have suffered breakdowns from, for example, failure to apply the concept of multipurpose planning, lack of funding, institutional weakness, and poor governance.

Among the nine countries in the basin of the world's longest river, the Nile, there are no agreements, nor is there any forum for negotiations on how its water should be shared. In the meantime, Ethiopia plans to divert 4×10^6 m^3 of water from the Nile every year for irrigation schemes, with serious consequences expected for Egypt and Sudan. Similarly, Zimbabwe plans to divert water from the Zambezi River, which it shares with Zambia, Angola, Botswana, and Mozambique, without any consultation. It is possible that a joint management plan, the Zambezi Action Convention, drawn up by the UN Environment Program (UNEP), will reduce the risk of conflict.

The risks of new tensions and conflicts, especially in North Africa, are clear, and the need for cooperation and agreement on the use of water resources has never been greater. The strategy, therefore, must be to recognize the role of cooperation in harmonizing nations' social- and economic-development strategies and instilling a sense of security among the member states. In the meantime, existing organizations must be strengthened, and new ones must be formed. Capacity-building and institutional arrangements are needed in these organizations to strengthen the planning units and improve policy and legal frameworks.

Community organizations and water supply

Although it is now accepted that community participation is essential for the success of any project, the water sector is still fraught with a heavy top-down approach. A central unit is essential for coordination of water policies, formulation of rules and regulations, and overall national planning, but this sometimes negatively affects the management of water resources at the grass-roots level. Ideally, implementation of projects, operation and maintenance of water activities, and management of water resources should be passed on to local communities.

Many African countries have completed or are on the verge of completing national water master plans, which will, through continuous updating, contain an inventory of existing water resources and their quality and quantity, including water supply, environmental problems, and rehabilitation needs. In addition, these countries have either national environmental action plans, national conservation strategies, or both to guide any analysis of water-related environmental issues. However, there are important barriers to community resource management.

The first of these barriers is the difficulty of devolving sufficient responsibilities for water management to local authorities and community institutions. Although a central policy-making body is required, water is extracted and used locally, so it is at the local level, primarily, where safeguards must be introduced. Decentralization, such as that undertaken in Botswana, needs to be planned on the basis of a complete drainage basin. One problem with a catchment approach is that the boundaries of the catchments and of the local authorities may not coincide. In many cases, various authorities will have to work together.

Another barrier to community resource management is the need to build the capacity of local authorities and local institutions to manage the resource. Trainees need "tool boxes" that describe visions and aims and explain how to set about planning and optimizing various parts of the development process. Regional and local decision-makers and officials also will need to be sensitized so that, with knowledge and vision, they will be inspired to act.

As physical planning and, to some extent, economic planning are being done at the regional–local level in many African countries, water-resource planning should also be carried out at that level. However, the local authorities may not have financial and jurisdictional powers over other sectors, such as agriculture, forestry, and mining; thus, the coordination of socioeconomic development required to maximize benefits from water-resource development is impaired. In addition, local authorities are known to lack financial resources to develop management issues fully. These barriers may be removed by clear water policies supported by encompassing laws.

To promote community participation, techniques that promote rational water use and management must be disseminated, and experience-sharing must be stimulated. Furthermore, research initiatives must seek to address the problems encountered by those who are affected and most concerned. The legal systems of most African countries need to recognize the various rural organizations and institutions that are ready to address local water needs and problems.

Water users' associations are new to Africa. However, with the liberalization of political and economic structures, resource-user associations are beginning to emerge, with the aim of managing their own local natural resources. The best examples of such user associations are the community wildlife-management groups in Zimbabwe and Kenya. In Kenya, the Lake Naivasha Riparian Owners' Association has been deliberating on land and water disputes since 1939. Presently, the association is involved in a three-phase study that will culminate in the formulation of an environmental-management plan.

Collaboration among communities, grass-roots institutions, and non-governmental organizations is essential for improving the management of fresh-water resources. Such collaboration also accelerates the sharing of experiences, knowledge, new ideas, and information between sector planners and practitioners.

Summary and recommendations

The availability of water resources in Africa is characterized by a striking paradox. On one hand, Africa contains many of the world's largest rivers and freshwater lakes, evoking a picture of an abundance of water resources. On the other hand, the Sahara and Kalahari deserts are surrounded by large tracts of marginal lands with rainfall of less than 700 mm/year. The overall water scarcity is, therefore, due to low runoff, high evaporation rates, and threats associated with nature (drought and salinity) and humans (overuse and pollution). The threats from human activities are most important because they can be controlled by improving water-management strategies. The impact of the environmental factors may be reduced by deliberate reorganization of socioeconomic policies.

Population stabilization

The rate of population growth in sub-Saharan Africa is very high, and, as a result, increased demand for water obliterates any gains that may have been achieved through increased focus and funding. Suggested methods for augmenting supply and managing demand will have little effect unless the rate of population growth is reduced considerably, for which the following are recommendations:

1. Governments need to mobilize both domestic and foreign resources to prepare for the inevitable increases in population that will further strain their freshwater systems. It is imperative that water conservation be combined with population strategies.

2. Conditions and strategies to help stabilize population growth and to achieve a sustainable balance between socioeconomic needs and available water resources should be instituted and encouraged. These include improved opportunities for women and better family-planning services.

Institution-building

Institutions and policies related to water-resource planning, development, and management are weak or absent in many sub-Saharan countries. In the past two decades, the problem of pollution from domestic, industrial, and agricultural sources has been growing. Industrial pollution is especially significant because of the absence of standards or a monitoring system. Several countries have problems with water loads contaminated by bacteria, organic matter, suspended solids, and nitrate pollutants. Expanding agriculture and associated increases in fertilizer and pesticide use are also a growing threat. Although many shallow groundwater resources appear to be contaminated by pathogenic agents, largely from fecal matter, there is no systematic water-quality monitoring system to regulate the problem. The negative impact of such uncontrolled contamination on the health of people is becoming more prevalent and visible, a problem for which the following are recommendations:

1. Water institutions must be strengthened and their mandates must be reviewed so that water-resource planning and development can be steered in an integrated manner.
2. An integrated and multisectoral approach to water-resource management needs to be established.

Capacity-building and technical assistance

Human-resource development is a major constraint to proper planning, development, and management of water resources. The following are recommendations to improve human-resource development:

1. Education and training programs on pollution control and hazardous-waste disposal need to be established, and existing ones need to be strengthened.
2. Countries should be given support to improve their capacity to identify and quantify problems and their causes and to find solutions, as well as to gather information and identify strategies for surveillance, equipment, and human-resource development.
3. Community-based management institutions, such as water users' associations, should be given support to manage freshwater resources.

Water policy and legislation

The following are recommendations to improve water policy and legislation:

1. National water-resource committees or commissions should be established to set policy and priorities.
2. Each country should establish a comprehensive national water policy.
3. Priorities in water policy, such as community water supply, health assurance and sound environment, cost recovery, and use of wastewater, must all be covered by law, to give them sufficient weight. Water law, where it exists, should be updated to deal with current management issues.

Water-resource planning and management

Water-resource assessment and monitoring are inadequate in many countries, and water standards are absent. Many countries have not used the data from IDWSSD and other projects in drought-stricken areas of Africa. In addition, excessive pumping of aquifers has led to negative economic and environmental repercussions, and some of the damage is permanent. An added constraint on availability of water also results from natural causes, such as high temperatures, high rates of evaporation, and decreased rainfall. The following are recommendations to improve water-resource planning and management:

1. Because each water source has a dynamic personality, general rules on water-source management are impractical. The responsibility for planning and managing water resources should, therefore, be decentralized and left to the communities and their institutions.
2. Safeguards on freshwater management must be introduced at the local level.

International cooperation

The following are recommendations for facilitating international cooperation:

1. International organizations, in cooperation with UNEP and WHO, must establish water-quality standards, provide workable guidelines, and promote water-source protection at all levels.
2. Risk-assessment methodologies and stress indicators need to be developed.
3. International cooperation and institutional capacities to monitor and control transboundary movements of hazardous wastes must be promoted and strengthened.

Demand management

The following is a recommendation for improving demand management:

1. Demand management, through water pricing, cost recovery, privatization, and community management, should be introduced.

Water-use efficiency and conservation

The following are recommendations to improve water-use efficiency and water conservation:

1. Strategies for improved efficiency and water conservation must be instituted.
2. Standards for water recycling and the use of water of marginal quality need to be established.
3. Modalities for reallocation of water rights must be developed.

4. External and internal sources of funding need to be mobilized, and appropriate use of water-generated public funds need to be ensured, with emphasis on transparency and accountability.

References

Ayoub, A.T. 1993. National soils policy for the protection and rational utilization of soil resources: technical, institutional and legal aspects. Paper presented at the 1st Crop Science Conference for East and Southern Africa, June 1993, Kampala, Uganda.

Burdon, D.J. 1977. Flow of fossil groundwater. Quarterly Journal of Engineering Geology, 10, 97–124.

Chilton, P.J.; Smith-Carington, A.K. 1984. Characteristics of the weathered basement aquifers in Malawi in relation to rural water supplies. Monograph.

Endersen, S.; Myhrstad, J.A. 1987. Multinational water schemes. Proceedings, International Water Symposium on Drought and Famine, July 1986. Olympic, London, UK.

Forster, S.F. 1993. Transboundary river basin management in southern Africa. *In* van Dam, J.C.; Wessel, J., ed., Transboundary riverbasin management and sustainable development. Proceedings, Vol. 1. Unesco/IHP, pp. 195–201.

Heidenreich, A. 1993. Pesticides in Bamako water: is it toxic waste? EcoNews Africa, 2, 12.

Khouri, N. 1989. Waste water reuse implementation in selected countries of the Middle East and North Africa. *In* Schiller, E.J., ed., Sustainable water resources management in arid countries: Middle East and North Africa. pp. 131–144.

Otieno, F.O.; Kilani, J.S. 1991. RBC or porous pots for textile waste treatment. Proceedings, 17th WEDC Conference on Infrastructure, Environment, Water and People.

UNECA (United Nations Economic Commission for Africa). 1989. Economic aspects of drinking water supply and sanitation in Africa with particular reference to rural areas. ECA/NRD/WRU/4/89, TPUB/9901.

UNSO (United Nations Sudano-Sahelian Organization). 1991. Assessment of desertification and drought in the Sudano-Sahelian region 1985–1991.

Wright, E.P. 1985. Water resources for humans, livestock and irrigation. *In* Geohydrology of drought prone areas of Africa. Proceedings of a seminar–planning meeting, Lobatse, Botswana.

Strain, Water Demand, and Supply Directions in the Most Stressed Water Systems of Eastern Africa

Maurice M. Ndege
Lecturer, Department of Civil Engineering, University of Nairobi, Nairobi, Kenya

Introduction

This paper examines sources of strain and water demand and supply directions in the most stressed systems of eastern Africa. According to hydrologists, an annual renewable freshwater availability per person of less than 1 000 m^3 constitutes water scarcity. This paper defines *stressed system* as a system where quantity and quality have been jeopardized because of overuse or exploitation.

The countries covered in this paper are Ethiopia, Kenya, Tanzania, and Uganda. Some attention will also focus on Somalia, Rwanda, and Burundi. Eastern Africa receives most rains from the monsoon system. The climate is equatorial, with great variation in the distribution of rains from the Indian Ocean front toward Central Africa and from the north to the south, as a result of different altitudes and latitudes.

Stressed water systems

The number of freshwater systems vary from one country to another in the region. In Burundi, fresh surface water is abundant, and springs are numerous. There are four hydrological regions: the Imbo Plains, the Zaire–Nile watershed, the high plateaus, and the Mosso Plains. The Imbo Plains include the hydrographic basin of the group of direct tributaries of Lake Tanganyika, including the Mushara, Rwaba, Murembwe, Dama, Ruzibazi, Mugere, Mutimbazi, and Ntahangwe rivers; and the Ruzizi Basin, which has 22 rivers, the main ones being the Luhwa, Myakagunda, Kaburantwa, Kagunuzi, and Mpanda.

All the rivers of Lake Tanganyika and the Ruvubu Basin rise in the Zaire–Nile watershed. The main watercourses of the high plateaus are the Ruvubu, Kanyaru, and Kagera. The Ruvyiranza is the southernmost course of the Nile. The Malagarazi flows into Lake Tanganyika after a long detour through Tanzania. In addition to Lake Tanzania, there are five smaller lakes: Cohoha, Rwera, Kanzigiri, Rwihinda, and Gacamirinda.

Ethiopia is well endowed with water resources. It has a number of Africa's major rivers and extensive drainage networks, which cover much of the highlands and considerable portions of the lowlands. There are 14 river basins, from which a total of 105.1 $\times$ 10^9 m^3/year flows out of the country into transboundary water: 78.7 $\times$ 10^9 m^3 to Sudan (77.5%), 16.1 $\times$ 10^9 m^3 to Kenya (15.9%), 6.5 $\times$ 10^9 m^3 to Somalia (6.2%), and the remaining 0.2 $\times$ 10^9 m^3 to the Red Sea. Most of the lakes are in the southern region.

The availability and suitability of the groundwater in the hard-rock formations vary greatly from one location to another, depending on the possibility of recharge, the density of the fractures, the permeability of the rock, the presence of internal obstacles to the movement of groundwater, the concentration and nature of the chemicals in the water, the depth of the aquifers' water table, and the physiography and difficulties encountered in drilling. The great depths and small yields of the boreholes are a hindrance to their wider use, other than for domestic purposes. Rational planning and exploitation of the resources are still limited because of inadequate reliable data and information about hydrogeological characteristics of various lithologic systems.

Kenya has five drainage basins: Lake Victoria, Rift Valley, Athi River, Tana River, and Ewaso Ng'iro. The country receives an average of 289.5 $\times$ 10^9 m^3 of rainwater annually, based on an estimated national annual mean precipitation of 510 mm. Most of this water escapes through evapotranspiration, and some infiltrates the ground. The rest is drained by rivers and streams into lakes and the Indian Ocean. Most streams are concentrated in the central highlands and western Kenya; in the rest of the country, dry valleys or seasonal rivers are common. Like Ethiopia, Kenya has an uneven distribution of water resources, which is well recorded. The Tana and Athi are the only two large rivers that transverse the dry areas in a southeastern direction, discharging into the Indian Ocean.

Groundwater resources are also unevenly distributed in quantity and quality. The drier areas have poor-quality water, with low borehole yields (less than 80 L/min), whereas the high-rainfall areas have many freshwater aquifers with high yields (about 117 L/min). In some areas, groundwater is not readily available; in others, potential aquifers lie at considerable depths and could be reached only through expensive drilling operations. In still other areas, available water is unsuitable for human consumption because of high salinity and high fluoride and other mineral-salt concentrations.

Rwanda has an abundance of surface water and springs. The main watercourses are the Kagera River and its tributary, the Nyabarongo. These are part of the upper Nile Basin and cover 40 450 and 14 600 km, respectively, with corresponding flows of 295 and 60–240 m^3/s.

Somalia, with the longest coastline in Africa, has two large rivers, which rise in Ethiopia: the Juba and the Shebeli. The Juba River crosses the country for more than 875 km; at Bardera the annual flow is 100–120 m^3/s, with a maximum of 1 100 m^3/s. The Shebeli is 750 km long and has a mean flow of 65 m^3/s at Beled Wyne, near the Ethiopian frontier, and 500 m^3/s at Afgoi–Andegle.

In Tanzania, the mean rainfall over most of the country varies from 250 to 1 000 mm. Higher rainfalls of 1 000 – 3 000 mm are recorded in northeastern Lake Victoria basin and in the southern highlands. Hydrologically, Tanzania is divided into five major drainage basins: the Indian Ocean basin, the internal drainage into Lake Eyasi, the Lake Natron – Buba depression complex, and the internal drainage complex.

The major river systems constitute the principal surface-water resources of the country, with mean annual runoff in millions of cubic metres. Half of the surface runoff flows into the Indian Ocean from the river systems of Pangani, Wami, Ruvu, Rufiji, Ruvuma, Mbwemkuru, and Matandu. The remainder drains northward, into Lake Victoria, westward, into Lake Tanganyika, and southward, into Lake Nyasa and the River Zambezi and then to the Indian Ocean. Some of the runoff also flows into drainage basins with no sea outlets. Groundwater is abundant in Tanzania and is a major source of water, particularly in the central regions of Shinyanga, Dodoma, Singida, and Arusha.

The chemical quality of water is generally good, except for pollution by municipal and industrial effluent of water sources in the Tanga, Kilimanjaro, and Arusha regions. Lake Victoria water is also polluted by untreated municipal sewage at a point close to the intake at the town of Mwanza. Salinity from brackish, saline connate waters or saline intrusion exacerbated by overpumping is found around Morogoro and in depressions near Lake Rukwa. High fluoride concentrations are found around the Arusha and Kilimanjaro regions. Boreholes are most common around the Dodoma, Singida, and Rukwa regions. The Ruvuma, Mbeya, and Kigoma regions have fewer than 100 of the country's 4 500 boreholes. Most of the water supplies come from boreholes.

Uganda is the "water tower" of eastern and Central Africa. The country has many high plateaus 900 – 1 500 m above sea level, and 18% of its area is covered with freshwater — swamps and lakes. There are seven main basins:

- Lake Edward (1 935 km^2) receives water from the Ntanga, Ishasha, and Nyamwere and other streams.
- Lake Victoria (58 161 km^2) receives water from the Kagera and Ruwizi rivers.

- Lake Albert (16 699 km^2) receives water from the Musizi, Nkusi, Wanbabye, Wake, and Weiga rivers.
- Victoria Nile (27 773 km^2) receives water the Kafu and swampy areas.
- Lake Kyoga (57 004 km^2) receives water from the Sezibwe, Victoria Nile, Okere, Okok, and Akweng. The water of this lake discharges into the Victoria Nile.
- Aswa is a tributary of the White Nile (26 558 km^2).
- Albert Nile (19 773 km^2) has tributaries from the West Nile hills.

Groundwater exploitation in the Karamajong region, an arid region, has yielded sufficient water to meet the needs of livestock and people. In areas around Lake Victoria, the groundwater has a salt content of 200–300 mg/L. Fecal pollution is a concern in Lake Kyoga and Karamajong, where groundwater aquifers rise almost to surface level during the rainy season.

Stressed systems occur in the northern districts, Jinja and Kampala regions, and around Lake Victoria. In Tanzania, the Rufiji, Pangani, Lake Victoria, Rukwa, and Ruvuma basins all are stressed. Kenya's stressed systems include lakes Victoria, Nakuru, and Naivasha and rivers Nzoia, Nyando, Turkwell, Kerio, Athi, Voi, Tana, and Ewaso Ng'iro. All of Ethiopia's rivers are under stress, except the Nile, which has not yet been used effectively for agriculture.

Sources of stress

A stressed water system is one that cannot adequately meet the demands of households, communities, and nations. The main factors that contribute to stress are population growth, irrigation, and livestock watering. Others include droughts and deforestation, poor land management, and pollution from human activities and industry. The most critical issue in the region is the deterioration of the water quality in lakes, rivers, springs, and groundwater, resulting in water resources becoming unfit for human consumption and other purposes.

Population growth

The annual population-growth rate of most countries in the region is 2.5–3.1%. This high growth rate, combined with economic development, results in ever-increasing demands for a finite resource. Hence, water availability per capita is steadily decreasing. Human activities have impacts not only on the water quality but also on the general availability of water resources and the state of aquatic ecosystems in the region.

Increased population pressure in large parts of the region has led to deforestation and increased cultivation. This, in turn, is affecting the hydrology and water balance and may lead to increased flood and drought problems, as well as to land degradation, soil erosion, and siltation problems. Afforestation may, however, lead to increased evapotranspiration losses and hence reduced water availability for downstream users as in the Ruaha Basin in Tanzania.

Irrigation

Tanzanian agriculture is mostly rainfed, and irrigation is used for protection against drought and for stable crop production. In some areas, communities use irrigation for dry-season farming, mainly growing vegetables. The national irrigation potential is 1×10^6 ha, 60% of which is in the Rufiji Basin. Two thirds of this potential could be used for double cropping. Irrigation potential is estimated by the availability and easy exploitation of water. The Rufiji Basin comprises the three major basins, the Great Ruaha, the Kilombero, and the Luwegu. About 60% of irrigation potential in the Rufiji Basin is in the Kilombero and lower Rufiji; 40%, in the Great Ruaha. The Pangani Basin has the best developed irrigated agriculture.

The critical situation affecting the users is the competition for water between agriculture and hydropower production. Because agricultural activity takes place upstream from Pangani Dam, any increase in farming will affect the level of water in the dam. In the Great Ruaha, more irrigation upstream from Mtera Dam will cause shortages in the dam, especially during drought. Because of this, the Tanzania Electricity Supply Company (TANESCO) insists on the closure of all irrigation systems upstream from the dam. Increased irrigation in the two basins will therefore have an environmental impact on the basin. At the national level, the government is trying to improve the water management of the basins by instituting prohibitive measures for users and punitive measures for offenders. It is likely that TANESCO will have increasing difficulty, because farmers will continue to irrigate their fields. There is a need for users to sit and discuss ways to equitably share this limited resource.

In Uganda, about 206×10^6 m^3 of water is used annually for irrigation. More than 32 510 ha is estimated to be under irrigation. Swamps provide the largest areas, with about 30 000 ha of small-scale irrigation in the Tororo, Iganga, and Pallisa districts.

There has been an increasing interest in rice cultivation by farmers. In areas surrounding the Doho rice scheme, hundreds of small-scale farmers grow rice outside the regular scheme; each farmer has an approximate cultivated area of 0.5 ha.

In the Lake Victoria crescent area, many horticultural farmers plan to start small-scale irrigation. However, the generally undulating topography precludes the use of gravity irrigation in most areas. This necessitates the pumping of irrigation water from lakes and streams, and some farmers have already bought electrical pumps for this purpose.

The Olweny Swamp Rice Irrigation Project aims to develop 800 ha of the swamp in Lira District for smallholder rice farming. Six hundred smallholders are the beneficiaries, with individual holdings of 1 ha. Inputs, credit, extension, and other services are provided under the project.

In 1992, the failure of the usually reliable first annual rainfall undermined the food security of Uganda, causing crop failures and decreases in livestock production. Of the 38 districts of Uganda, 15 experienced a long dry spell; food-security problems reached crisis levels. The areas worst affected were Kasese, Kabale, Mbarara, Rakai, Bundibugyo, Masaka, Masindi, Mpigi, Mukono, Luwero, Moroto, Kumi, Soroti, Kotido, and Rukugiri.

In Kenya, irrigation is carried out at both the local and national levels. People are aware of the potential boost in harvest from irrigation. The present national policy relating to irrigation-water supply emphasizes taking low-cost approaches to implementation while increasing the acreage under irrigation.

The government has instituted the National Irrigation Board, and big irrigation schemes like the Mwea, Bura, and Ahero are partly government controlled. But unless legislation is enforced, communities upstream may use all the water, leaving little for downstream users. Another pressing issue of a regional nature is that extensive irrigation from rivers that flow into Lake Victoria may interfere with the ecosystem. The lake level will fall, and more pollutants will find their way into water. The Water Appointment Board, the body legally empowered to control and regulate the abstraction of water (surface water and groundwater) needs strengthening. Currently, the *Water Act,* first enacted in 1962 and revised in 1972, is awaiting another revision and approval by Parliament. In terms of protecting water resources from pollution the Act is weak. In a move to avert any environmental damage associated with flood irrigation projects, environmental-impact assessments are now preconditions for starting any irrigation projects in Kenya. In Ethiopia, Somalia, Rwanda, and Burundi, irrigation is not well developed.

Livestock watering

Livestock makes significant water demands, especially in the semi-arid, pastoral areas, where surface-water sources are scarce and long dry seasons are experienced. The seminomadic pastoralists who inhabit these areas often encroach on natural reserves, such as Lake Mburo National Park in Uganda, in search of water and pasture. In the past, 425 medium-sized dams and valley tanks, as well as several small valley tanks, were provided. Most of these are silted because of lack of maintenance, poor animal-watering methods, and soil erosion that results from overstocking.

Water use

Domestic and industrial use

Tanzania

In Tanzania, most of the domestic and industrial water supplies are from surface water. Groundwater sources, though potable in most cases, are not used because abstraction requires sophisticated and sometimes expensive technology.

The installed capacity for rural water-supply schemes, as of June 1992, served about 47% of the population. However, the reliability of the data is questionable because more than 35% of the schemes were not in operation. Many of the pumping units were worn out and nonoperational and needed replacement. Urban water supply, by June 1992, served about 67% of the population. This figure didn't take into account the quality of water supplied. Sometimes, because of the nonfunctioning of treatment plants and nonavailability of water-treatment chemicals, water is supplied either partially treated or untreated. The operational costs are normally higher than the revenue collected. This is because water tariffs do not meet running costs, and billing and revenue collection systems are insufficiently streamlined. In both the rural and the urban sectors, water demand far exceeds supply. Table 1 shows the low water-supply coverage.

Table 1. Status of Tanzania's domestic water-supply sector (1992).

	Population (million)	Population-coverage target		Actual coverage [a] (%)
		(L/d per capita)	(%)	
Rural	15	25	90	47
Urban	1.7	50	10	67

[a] Reliability of the data is questionable because more than 35% of the schemes were not in operation.

Table 2 shows the distribution between investments in the rural and urban water supplies. The apparent bias in investment, toward the urban centres is attributed to differences in the levels of service, technology, and institutional requirements in the two areas. For example, although per capita investment in the rural areas may be as low as 6 United States dollars (USD) (spring protection), the corresponding figure in the urban centres often exceeds 120 USD.

Table 2. Planned investment in Tanzania's domestic water-supply sector.

	Coverage (%)		Investment (USD/m³)	
Year	Rural	Urban	Rural	Urban
1992	26	60	14	30
1995	36	75	30	40
2000	75	100	40	40

Uganda

In Uganda, water supply and sanitation in the urban areas are provided by the National Water and Sewerage Corporation (NWSC) and the Department of Water Development (DWD). NWSC is responsible for supplying water to about 1 million people in Kampala, Entebbe, Jinja, Mbale, Tororo, Masaka, and Mbarara. The average coverage is 51% of the urban target population.

Existing water-supply systems are in a poor state of repair because of maintenance constraints, and most often the population has to rely on unsafe water sources. An umbrella program, the Rural Towns Water and Sanitation Programme, has been instituted to coordinate all the urban water projects under DWD. This reflects a major shift in government policy toward decentralization and represents a demand-driven participatory approach.

In 1992 DWD estimated the rural potable-water-supply coverage to be 26%. The level of investment required to raise the coverage to 100% is estimated to be 351 million USD. The major rural water-supply development programs are RUAWASA East Uganda Project (financed by Danish International Development Assistance), running up to 2000 and covering eight districts; and SWIP (financed by United Nations International Children's Emergency Fund [UNICEF], Canadian International Development Agency, and Swedish International Development Agency), covering nine districts. Other programs include WATSON the National Water and Sanitation Programme (financed by UNICEF and various nongovernmental organizations [NGOs]), covering nine districts; and the West Nile Rural Water Supply Programme (financed by Italy and NGOs), covering two districts.

Ugandan industry is mainly engaged in processing raw materials from agriculture, livestock, and forestry. Major industrial activities include the production of textiles and garments, leather, sugar, foods, soft drinks, beer, and flour. These activities are concentrated in southern Uganda, particularly Kampala and Jinja, on the shores of Lake Victoria and the Victoria Nile. Uganda had a strong industrial base in the 1960s, but this was destroyed during the 1970s. To date, there are only about 5 000 factories, many of them operating below capacity. Industry contributes 5% of the gross domestic product, and industries are generally connected to the urban water-supply networks.

Kenya

In Kenya, water for agricultural use will continue to command the highest demand. It is projected that the national water demand will progressively increase from 5.68×10^6 m^3/d in 1990 to 15.94×10^6 m^3/d by 2010. Of this, 73% will be for agriculture, 4% for livestock development, 22% for domestic use and industry, and 1% for inland fisheries and wildlife.

The government has recognized the need for environmental protection and promoted various measures to ensure sustainable agriculture. These include intercropping, under the agroforestry program, soil and water conservation measures, and proper use of agrochemicals to minimize adverse environmental effects. Other measures will include a shift from agrochemical dependence to organic farming, which has a lower environmental cost.

The water-supply needs of a rural person are estimated at 50 L/d; those of the urban dweller, 100 L/d. According to estimates, the total rural water-supply demand, based on the above figures, will be 749.3×10^6 L/d by 2000; the total urban water-supply demand, 1.17×10^9 L/d. The water demand for wildlife has been estimated at 21.0 m^3 of fresh water per day, on the basis of wildlife species and their distribution in the country. Water for fish-farming needs, based on a fish-production capacity of 2.5/ha, has been assessed at 0.96 m^3/s.

Hydropower

In Uganda, most hydropower generation takes place at Owen Falls Dam, at the Victoria Nile near Jinja. The present installed capacity is 180 MW. An extension program intended to bring about increases of 270 and 300 MW of generating capacity has been prepared. The identified hydropower potential on the Victoria Nile within the Ugandan territory is 2 700 MW, with the Murchison site having a potential for 600 MW and the Bujagali site having a potential for 250 MW.

At present, there is no artificial storage on the upper Nile, and river flows have been unchanged by the construction of the Owen Falls dam. The mini-hydropower stations have small storage reservoirs, and there is a slight tendency to equalize the natural river flows.

Hydropower generation can be expected to increase as a result of the transboundary distribution of electricity and the increase in rural electrification and industrial and in domestic demand. However, the impact on the water resource (Victoria Nile) will be negligible.

Most of Tanzania's hydropower potential is in the Rufiji River system. Other rivers with hydropower potential are the Kagera, Ruhuhu, Wmai, and Rufirio. Most of the hydropower potential of the Pangani and Great Ruaha has been developed. Of the existing power-generating facilities, 86.5% are hydropower units. Of the total available hydropower-generation capacity, more than 99% is in the Great Ruaha and the Pangani.

In Kenya, about six hydropower projects are considered promising. These are proposed for commissioning toward 2010. There are also plans to develop hydropower on the Yala and Nzoia rivers.

Responses to water stress

Governments' response

Governments are recognizing that the problems of water stress are not confined solely within their national borders. Issues related to the management of international waters are being addressed at various forums, and various regional bodies have been formed for such purposes. One example is TECONILE, a body incorporating all the countries of the Nile Basin. TECONILE was formed for the sole purpose of proper management of the basin's water resources. Treaties signed years ago are being revised and drafted to take into consideration the needs of all the upstream and downstream users.

Governments are also undertaking the following:

- forming high-level, cross-sectoral water-policy committees to formulate guidelines for task forces and to coordinate the legal framework for management of shared water resources;
- setting up modalities for tariff charges and prices;
- embarking on national environmental programs to preserve and conserve forests;
- promoting the operation of water projects by entrepreneurs in rural areas and by autonomous water entities in urban centres;
- preparing guidelines for estimating water-related opportunity and environmental costs; and

- preparing dynamic water action plans (covering water-resource assessment, required institutions, management instruments, etc.) based on water-resource policies.

Uganda, Tanzania, and Ethiopia are preparing water action plans, but Kenya has yet to develop one.

Communities' response

Communities are undertaking the following to address the problem of water stress:

- adopting technologies and management approaches that increase the efficiency of water use, allocation, and distribution (such technologies and management practices make it easier to conserve water, to increase the efficiency of water use and conveyance, and to reuse wastewater);
- drilling boreholes, shallow wells, and pit and VIP latrines;
- discussing, planning and implementing development projects in their areas;
- promoting environmental protection practices, like planting trees and establishing nurseries;
- creating intersectoral project task forces;
- organizing education programs, targeted at the household level, on the benefits of potable water and on related costs, especially for operation and management; and
- setting prices and collecting fees to cover operation and management, either in kind or in cash.

Conclusion

Sources of stress should be looked at, not just from the hydrologic point of view, but also from the perspective of accessibility (distances users have to travel to collect water), socioeconomic development, poverty, pollution, human and livestock population increases, and increases in agricultural production (for instance, irrigation acreage).

Ethiopia, despite all its rainfall, has access to only 9% of its water resources for development because 91% flows into international waters. Each country should develop water policies, to be followed by water action plans. Uganda is a good example to be emulated.

Research opportunities

Research opportunities and needs for different countries vary and may be linked to the levels of development and economic stability of each country. The following research initiatives are recommended.

Kenya

1. Launch a national exploratory program to map out areas suitable for future water development; identify the most suitable and economical methods for such development.
2. Address pollution laws covering the contamination of surface waters and groundwaters by various activities and make the laws practicable and workable.
3. Come up with pollution indicators that ordinary people can use at the community level.

Tanzania

1. Adopt the technology needed for studying and recovering groundwater.
2. Address the pollution from municipalities and agricultural sectors.

Uganda

1. Identify sustainable sources of water for human beings and livestock.
2. Use technical staff from water-related government departments, instead of the private sector, to constrain water schemes. Look into the effects of this bias toward using the private sector on human-resource development at the national level.
3. Look at pollution bylaws and improve them for efficient management of the water sector. In cases where no bylaws exist, develop some and enforce them.

Governments of all countries in the region

1. Study hydrological regimes to understand why some rivers, like the Katonga and Kafu, in Uganda, are drying up.

2. Develop technology for studying and recovering groundwater, including that in contaminated wells.

3. Study the problems related to the operation and maintenance of water-supply schemes, with a view to making them sustainable and operational.

4. Develop mechanisms, tools, and models for updating information in the water sector.

5. Examine country-level institutions.

6. Work out a cooperation framework for the management of common water resources.

STRAIN, WATER DEMAND, AND SUPPLY DIRECTION IN THE MOST STRESSED WATER SYSTEMS OF LESOTHO, NAMIBIA, SOUTH AFRICA, AND SWAZILAND

Kathy Eales, Simon Forster, and Lusekelo Du Mhango
Consultants, Economic Project Evaluation (Pty) Ltd, Rivonia, South Africa

Introduction

This paper addresses the sources of stress (including conflicts among uses and users of water) and the effects of this stress on the availability, accessibility, and quality of water in Lesotho, Namibia, South Africa, and Swaziland. In addition, water-demand–management strategies aimed at resource conservation and alternative water-supply sources and mechanisms are considered.

The approach taken in this study was to collate all available literature on the water situation in each country and then to follow up with visits to the countries to interview government staff, academics, consultants, and nongovernmental organizations (NGOs). To preserve the confidentiality of the discussions, we do not identify the sources of the views and opinions presented. Data were collected mainly from government sources, with some inputs from consultants.

Lesotho

Water supply and demand

Water availability

The surface-water resources of Lesotho are substantial and far exceed the present and future needs of the nation. However, the high runoff is often rapid and occurs in inaccessible mountainous terrain. Major capital-intensive engineering works (unaffordable to Lesotho) would be required to harness this water for use by people.

Less than 9% of the land is arable. There is an acute shortage of land for settlement, overgrazing is severe, little fuelwood remains, and the annual rate of population growth is 2.6–2.9%. The result in many areas is acute environmental degradation, manifested by soil erosion and silting of rivers and dams. One reason for this degradation is that the Basotho people have only lived in the area for the past 140 years, and although their stock-management practices were well suited to the sparse settlement patterns of the flat Orange Free State and Transvaal, these practices have not been sufficiently adapted to the mountainous terrain of Lesotho.

The sedimentation of Lesotho's surface-water resources has serious implications for South Africa and Namibia as well, because the headwaters of the largest drainage system in South Africa, the Orange River, are located in Lesotho. About half the total flow of the Orange River is provided by the Senqu River.

Despite the high availability of surface water, there are problems with its suitability for community water supply. Untreated surface water is generally not considered potable because of Lesotho's high livestock levels, and water is rarely boiled because of the scarcity of fuelwood.

Although the total potential yield of Lesotho's groundwater resources is unknown, this source currently accounts for about 70% of the water used in Lesotho and is derived primarily from a large number of low-yield springs. People in the highlands use mainly spring water, whereas in the lowlands the rural communities are provided with water from both springs and boreholes. Groundwater availability is erratic, and aquifers are generally discontinuous along doleritic dikes. This makes drilling of boreholes unpredictable and expensive. Most boreholes yield less than 0.5 L/s. Recharge is extremely slow, and boreholes are frequently pumped dry.

Water demand

Water demand in Lesotho centres primarily around domestic and stock-watering needs. Annual demand is projected to increase from 26.7×10^6 m^3 in 1994 to 46.2×10^6 m^3 in 2020. Most towns rely on a mix of surface water and groundwater, although groundwater represents only 10% of total urban water consumption. Two urban settlements, Morija and Mapoteng, with a combined population of about 9 000, rely solely on groundwater.

There has been no long-term monitoring of changes in the yield of springs, but anecdotal evidence suggests many springs are drying up because of the loss of protective vegetation and soil. In turn, this loss promotes rapid surface runoff, as opposed to infiltration.

Water-management systems

Lesotho's Department of Water Affairs is responsible for hydrological services, resource management, development of a water-resource master plan, and water-quality monitoring. Coordination with other agencies addressing water resources is poor. Management of soil erosion falls under the Department of Agriculture. Rural water supply falls under the Village Water Supply Section (VWSS) of the Ministry of the Interior; urban water supply, under a commercialized parastatal. No master plan or national water-supply strategy exists.

Rural water supply

Of Lesotho's 1.8 million people, 1.53 million (85%) live in rural settlements. Of these, 890 000 (58%) have access to improved water-supply systems, and the remaining 640 000 draw water from unprotected rivers, springs, and earth dams.

Rural water supply is the responsibility of VWSS. VWSS operates at three main levels: district, regional, and national. Communication among these levels is generally good, but with other government departments it is unstructured and informal and, between VWSS and villagers, limited. There is very little evaluation of projects, either before construction commences or after commissioning.

Government officials generally have less faith in the success of local water management than do the NGOs working in close contact with village committees. The head of VWSS believes that voluntary community structures don't work without cash incentives and maintains that if the government doesn't provide all the necessary resources, there will be a further breakdown. However, the government lacks the staff and resources to take full control of rural water supply. In contrast to this view, a recent report by consultants indicated that the rural water-supply system is starting to work relatively well, with local villagers managing basic maintenance and repairs fairly competently. Table 1, which is based on a mid-1994 survey of the Maseru region (regarded as typical of the mountainous areas), illustrates this point.

Water-improvement projects in Lesotho's rural areas are initiated at the village level. Villagers are required to form a village water committee (VWC) and start collecting funds toward construction and future maintenance. They approach VWSS through village leadership structures; in due course (this can take several years), VWSS provides a technical team of masons, engineers, and whatever else is needed. The major task of the VWC is to coordinate the inputs of local residents during the construction and installation of an improved water system. This includes free labour and accommodation for VWSS personnel. A major effort is required from local residents: trenches, often many kilometres long, have to be dug to lay pipes, and stones have to be cut for the construction of siltboxes and water tanks. Average construction time is 170 d, but some schemes can take more than 2 years to complete. Women carry out the bulk of this work because many men are absent as migrant labourers and because women tend to be the primary

Table 1. Survey findings from 209 water projects in the Maseru region of Lesotho.

	Proportion (%)
Systems that are handpumps	36
Systems that are gravity fed from springs	27
Systems that are a combination of handpumps and protected springs	37
Villages that have fully functional water-collection points	75
Villages that have a VWC	90
VWCs that have a bank account for repairs	69
VWCs that have a tool box	33
Villages that have a water minder	49
Villages that have a water minder who can undertake repairs	29

Note: VWC, village water committee.

beneficiaries of water schemes, through reduced time spent collecting water and reduced risk of family illness from contaminated water.

Once the project is complete, the VWC tends to become inactive until a problem arises. Residents are supposed to try to fix the problem themselves, and when they can't, they notify the district VWSS office. In due course, a technician is sent out. VWSS can barely cope with breakdown maintenance and does not undertake preventive maintenance. VWSS tries to recover 50% of the cost of repair from local residents, in instalments over 3 years if necessary. It seldom achieves this.

The distinction between VWCs and village development committees (VDCs) is breaking down. In the past, VDCs were regarded as the local arm of the ruling political party, and thus separate water committees were introduced in the 1970s to ensure that water development was not derailed by political posturing. This need for separate structures seems to have fallen away now, and there is talk of merging the two bodies.

Mistakes are inevitable in any rapid-development initiative. Lesotho's most serious mistake seems to be that in VWSS's eagerness to meet water demand and its own goals, it installed new water schemes faster than its ability to service them or train users to maintain and manage them. Its aim has been to achieve full coverage of rural areas by 2005, and it is well ahead of schedule. However, the sustainability of these systems is questionable. Maintenance is a major problem, and it is estimated that only 40% of Lesotho's boreholes are operational.

Since 1992 VWSS has been trying to reformulate policy, determine strategies, and collect supporting data. The first regional report on coverage, condition of water systems, demographic trends, and local organizational capacity was completed in August 1994. Field inspections are now generating the kind of detailed data VWSS needs to begin to assess current capacity and evaluate strengths and shortcomings.

Urban water supply

The Water and Sewerage Authority (WASA), a parastatal set up by the Ministry of Natural Resources, relies on surface water for 90% of the water it supplies to urban settlements. The largest urban settlement is Maseru, with a population of 90 000. The next largest town in Lesotho has a population of less than 10 000. Most urban water is taken directly from local rivers and stored in reservoirs. However, the silting of both the rivers and the reservoirs supplying urban areas is a major problem.

Although the coverage of the water supply is adequate, WASA has management problems. It needs to spend more on upgrading its infrastructure to reduce water wastage and losses. Moreover, because of high connection costs for individual households, many people, most of whom could afford to pay regularly for water if the connection fee was lower, are drawing water free from public standpipes.

The Lesotho Highlands Water Project

If fully developed, the Lesotho Highlands Water Project (LHWP) could see the diversion of 2.2×10^9 m^3 of water per year from the headwaters of the Orange River to the Vaal River, making it available under gravity to Gauteng Province, South Africa's industrial heartland. The scheme is highly capital intensive.

Although Lesotho residents will not be supplied with water from the two dams that will be built as part of the LHWP, there are significant indirect benefits. For example, the development of infrastructure includes rural roads, health centres, schools, and a few village water-supply systems. There are also the royalties from the diversion of the water, self-sufficiency in electricity generation, and revenue from Lesotho's participation in the Southern African Customs Union Pool, which has received a boost from the importation of construction equipment and materials. The indirect costs of the scheme will be some environmental damage and the loss of some agricultural land through inundation.

Given the acute shortage of arable land in Lesotho, the loss of even a small percentage of agricultural land has serious implications for subsistence agriculture. Elaborate and sometimes controversial schemes have been instituted to compensate farmers for the loss of productive land.

Some doubt has, therefore, been cast on the merits of the LHWP. If it is fully completed, water transfers from the Orange River's headwaters into the Vaal River could jeopardize the assurance of supply to South African irrigation schemes lower down in the Orange River and in the Fish–Sundays catchment of the Eastern Cape, which receives water transferred from the Orange River. However, the benefits of an assured water supply to Gauteng are believed to exceed the negative impacts.

Major constraints and recommended research

Erosion and silting

PROBLEM — Overgrazing on the fragile soils of Lesotho's steep slopes is causing serious sedimentation of rivers and water-supply reservoirs, as well as reduced recharge of water-supply aquifers.

RESPONSE — Piecemeal soil conservation techniques are employed by the Department of Agriculture, but these do not address the complex underlying socio-economic problem of overgrazing and dense settlement on vulnerable soils.

RESEARCH — A multidisciplinary approach to soil conservation, focusing on community-based socioeconomic incentives, needs to be investigated.

Rural water supply

PROBLEM — Of rural households, 43% do not have access to clean water; most of these people live in remote areas. VWSS has overemphasized installation and construction of boreholes at the expense of developing maintenance capacity. Although VWSS has only a limited capacity to repair breakdowns, there are no private-sector borehole-repair teams. In addition, village water minders do not always have the skills they require to do preventive maintenance on handpumps.

RESPONSE — Despite a profound awareness of the problem, VWSS does not have the capacity to remedy the situation.

RESEARCH — Investigations need to be carried out to ascertain the best ways to ensure both the adequate transfer of maintenance skills to local people and the retention and effective use of these skills by the community.

Coordination and planning

PROBLEM — No overall picture of national demand and supply exists, water management objectives are imprecise, and interagency communication is poor.

RESPONSE — Academics at the University of Lesotho have been commissioned to investigate demographic trends and projected urban water demand. This will feed into a Water Resources Action Plan project, which began in December 1994. Lesotho urgently needs a water-management master plan and is currently looking for funds to commission this.

Legislation

PROBLEM — Lesotho's *Water Act* of 1978 has deficiencies and is largely ignored. It does not provide government agencies with the powers they need to resolve water disputes, nor does it provide for effective resource management and pollution control.

RESPONSE — The *Water Act* is being revised, but no completion date has been set.

Revenue

PROBLEM — The revenue from water connections and sales in urban areas needs to be increased if infrastructure upgrades are to be self-funding.

RESPONSE — WASA has recruited (with overseas assistance) the necessary skills to review tariff policies. Whether this will lead to a reduction in connection fees to ensure more paying customers is uncertain.

Namibia

Water supply and demand

Water availability

Namibia is the driest country in sub-Saharan Africa. It is estimated that 83% of all rain evaporates soon after it falls, leaving just 17% available as surface runoff. Of this runoff, 1% recharges groundwater sources, and 14% is lost through evapotranspiration. Only 2% of the total rainfall can be captured by surface-water-storage facilities.

There are no perennial rivers within Namibia, only ephemeral ones. Perennial rivers are found on Namibia's borders: the Orange in the south; the Cunene in the northwest; and the Okavango, Kwando-Chobe-Linyati, and Zambezi in the northeast. Namibia currently has access to an agreed 180×10^6 m^3/year from the Cunene River and at least 500×10^6 m^3/year from the Orange River. No formal agreements have yet been reached on abstracting water from the Okavango River. However, the completion of the last stage of the Eastern National Water Carrier, the largest state water project in Namibia, will lead to the importation of 100×10^6 m^3/year from the Okavango River to augment supplies to the central, eastern, and western areas of the country.

The flow in the ephemeral rivers in the interior is irregular and unreliable, limiting both the potential for utilizing surface-water sources and the recharge of aquifers from river courses. The estimated safe yield of the surface-water works that could be developed on the ephemeral rivers is at least 200×10^6 m^3/year, or 40% of the total surface-water resources available in the interior. Ten large dams have been constructed on these ephemeral rivers, with a combined safe yield of 87.3×10^6 m^3/year.

Groundwater plays a major role in water supply in Namibia. The safe annual yield from groundwater sources is estimated at 300×10^6 m^3/year. However, overabstraction of groundwater is already a serious problem in some areas. In the karst (limestone) areas, excessive pumping from boreholes can result in the deeper lime-rich water being exposed to oxygen and thereby causing the lime to precipitate and block the borehole. The borehole then has to be abandoned or redrilled.

A more serious problem is the depletion of the aquifer itself. There are various examples that illustrate this issue:

- The Kuiseb River alluvial aquifer in the central Namib area has already been overused, and the water table has dropped significantly. The aquifer can no longer meet the needs of the coastal towns of Swakopmund and Walvis Bay or of the Rossing uranium mine, and the lowered water table has seriously undermined the dependence of the local Topnaar people on hand-dug wells for water.
- In the Kuiseb and Omaruru catchments, the combination of bad farming practices and prolonged droughts has reduced the vegetation cover, leading to considerable topsoil removal during intense rainstorms and the subsequent sedimentation of the Kuiseb and Omaruru rivers. This soil often forms a thin layer of fine material on the riverbed, which seals the surface of the sand and prevents groundwater recharge.
- The fossil water from the aquifer under the Koichab River is being mined to support the town of Luderitz and the industry at Elizabeth Bay. The aquifer will probably never be recharged under present climatic conditions.

- In some parts of the Stampriet artesian aquifer, saline water overlies the freshwater and poses a contamination threat to the freshwater. Farmers in this area are now required to use a specially designed borehole that seals off the overlying salty water.

Thus, in many areas, the abstraction of groundwater and the impoundment of surface water have upset the delicate balance sustaining highly vulnerable ecosystems.

Water supply

Namibia's total population is about 1.5 million. About 1 million people, or 65% of the population, live in the underdeveloped northern regions, mostly in rural settlements. The northern population is further concentrated in the centre of what used to be called Owamboland. Here 400 000 people live in an ephemeral wetland system of pans called the *oshanas*. Much of the groundwater in this area is too saline for human consumption. In addition, because of the extreme aridity of much of Namibia, most rivers and aquifers within the country may be regarded as under stress. However, extensive investments in pipelines, canals, interbasin transfers, and improved abstraction technology have made relatively dense human settlements viable in areas that previously could not have supported close settlement on this scale.

The *oshanas* run from north to south, whereas the pipelines, canals, and main roads run from west to east, obstructing the *oshanas*' normal flow and increasing evaporative loss. This, in turn, reduces groundwater recharge, which then threatens the water supply in those traditional settlements not serviced by the pipelines and canals. One quarter of Namibia's population live in the *oshana* area, and this is expected to double in the next 20 years.

Depending on the water region, rural people draw water directly from rivers and natural springs, dig for water in dry riverbeds, or use hand-dug wells.

In the *oshana* water region, groundwater takes two forms: a deep saline aquifer underlies most of the area; above this are perched aquifers, pockets of freshwater (rainwater) trapped between the surface and the saline water. Rural people dig into the fresh groundwater aquifer by hand, making round wells called *omifimas*. Water is abstracted with buckets.

Surface water is available during the summer months, when there is sufficient rainfall to make the *omifimas* flow and fill hand-dug earth dams. Initially, the quality of the fresh surface water is good, but as the water evaporates its salinity increases. Traditionally, rural communities would then move on to other water sources in the dry months. Rapid population growth, dense settlement, and environmental degradation are making this migratory lifestyle more difficult.

The United Nations Development Programme planned to set up pilot schemes in 1995 to investigate combining traditional hand-dug wells with infiltration galleries to provide low-technology filtration systems.

Nonconventional water sources

DESALINATION — A pilot desalination plant using seawater is being established on the west coast of Namibia, but the cost will be high, roughly 6.50 ZAR/m^3 (in 1996, 4.34 Namibian dollars [ZAR] = 1 United States dollar [USD]), including capital costs, if used locally. Pumping desalinated water inland to Windhoek is not economically feasible at present because of prohibitive pumping costs.

Most of the groundwater in the far north is saline. Desalination schemes for groundwater have been tested, but major investment in desalination schemes is unlikely because the groundwater is a limited resource.

Water recycling is practiced in some areas. Windhoek recycles about 12% of its water, and the Rossing uranium mine recycles about 76%. However, it is widely acknowledged that recycling could be greatly improved in other centres.

ASSISTED RECHARGE — To conserve Omaruru River floodwater from being lost to the sea, a dam has been built on the lower Omaruru River, near Walvis Bay, to trap silt-laden water during floods. The silt settles out behind the dam wall, and the clear water is then pumped downstream to sand-filled basins, where it rapidly infiltrates, recharging the aquifer. This water is later pumped out via boreholes. Earth dams are used for assisted recharge in some areas, but high evaporation rates reduce their effectiveness.

DROUGHT MANAGEMENT — Given Namibia's aridity, drought should not be regarded as exceptional. However, before independence, water tankers were extensively used in rural areas during droughts. This practice was revived during the 1992 Emergency Drought Relief Program, at a cost of about 50 ZAR/m^3, to augment other relief measures. However, given the logistics and cost of transporting water over long distances, the new government considered this an inappropriate response. Current drought-relief strategies focus on improving borehole reliability.

OTHER OPTIONS — Large-scale rainwater harvesting, weather modifications, and fog-harvesting systems have been investigated by the government but rejected, as they were shown to be uneconomic. Of course, rainwater harvesting at household level can prove feasible and is often practiced in areas where groundwater has a high salt content and where precipitation is sufficient.

Water demand

Groundwater meets 57% of Namibia's current water demand, and surface water meets the remainder. Settlements in the far north are aggregating along the network of pipelines and canals that connect the major rural villages with the Cunene River and supply water to about 30% of the northern population. Livestock are believed to account for 80% of all water demand in northern Namibia; much of this demanded is being met by the pipeline. Moreover, livestock that previously were moved from one water point to another are now settled in fixed areas, leading to overgrazing and overstocking. Fixed human settlement is also denuding vegetation as trees and bushes are used for fuel and buildings.

Because Namibia relies heavily on major interbasin water-transfer schemes, demand statistics expressed in terms of surface-water catchment are not particularly useful. Furthermore, aquifers are seldom contiguous with catchment boundaries and can even be subject to interregional transfer, as in the case of the Karstveld area around Grootfontein, where groundwater is exported southward to Windhoek. Therefore, sectoral demand is best compared with water availability on the basis of existing abstraction patterns, as shown in Table 2, which shows current demands and demands projected to 2005.

Table 2. Current and projected demand in Namibia by sector.

	Demand ($\times 10^6$ m^3)					
	Perennial rivers		Ephemeral surface water		Groundwater	
Sector	1990	2005	1990	2005	1990	2005
Urban [a]	12.6	40.0	13.4	45.0	41.0	45.0
Irrigation	39.7	95.0	34.8	30.0	31.5	30.0
Stock	3.7	10.0	0.0	65.0	60.3	65.0
Mining	2.0	10.0	2.5	10.0	7.5	10.0
Tourism and environment	0.0	2.0	0.3	2.0	0.7	2.0

[a] Includes domestic and industrial demand.

Only 49% of total estimated ephemeral surface- and groundwater sources will be used by 2005, but demand on perennial rivers is expected to increase by 270%. Irrigation demand is unlikely to increase dramatically because of generally poor soil quality. Any additional irrigation demand will probably be met from the perennial border rivers.

Reconciling future supply and demand

Unfortunately, much of the potential water resources available to Namibia are not located close to where they are required. For example, there is abundant water in the Fish River, but it is far from any human settlement. Therefore, as the surface drainage system experiences such high losses, the future utilization of perennial rivers will entail large, capital-intensive engineering projects. However, such schemes may not be affordable or internationally acceptable. Moreover, the diversion of water from internationally shared perennial rivers will require extensive negotiations with Namibia's neighbours.

Water-management systems

Rural water supply

Despite significant problems, rural domestic water-supply coverage is generally good (around 55%) and well within the United Nations International Children's Emergency Fund target for 2000. However, it is estimated that at least half the existing water points in rural areas are faulty.

Responsibility for rural water administration in Namibia has been reassigned three times since independence in 1990. Before independence, rural water supply was the responsibility of the eight ethnic regional administrations, with limited funding and capacity. The result was a massive backlog of communities with inadequate water supplies. Failure to involve and train local users led to a high number of system breakdowns, which was compounded further by the lack of routine-maintenance capacity during the war of liberation.

In early 1993, responsibility for rural water supply was transferred to the Directorate of Rural Water Supply (DRWS) in the Namibian Department of Water Affairs (DWA). With external assistance, a new model of water administration, designed to change the role of DRWS from that of provider to that of facilitator, is being introduced. Because of the lack of capacity within rural communities, the role of DRWS extends well beyond facilitation. Much of the work of DRWS focuses on developing institutional capacity. However, at this stage, only half the posts in DRWS are filled, and few of the existing personnel are appropriately trained. The success of DRWS depends on whether it can recruit and train sufficient staff members and enlist the cooperation of the rural communities with which it works.

Namibia has been divided into 10 water-supply regions, each with a chain of water committees from local water-supply points up to district level. DRWS asks communities to sign a contract giving the communities ownership of their local infrastructure and requiring them to undertake and fund routine maintenance. Each water committee will have a caretaker trained in preventive maintenance. For every 20 or so water-supply points (depending on distances between points or terrain), there will be one rural water extension officer resident in the area. This person will be able to summon help for breakdowns and maintenance from DRWS's regional maintenance section. On paper, the scheme looks impressive, but it has not yet been implemented widely, so it is too soon for a critical evaluation.

With major donor and NGO assistance, DRWS has prepared a range of educational materials to improve understanding of the hydrological cycle and of the importance of appropriate resource and stock management. Booklets have been designed to assist in promoting literacy and to be used in conjunction with radio broadcasts. Caretaker manuals for diesel and handpumps, with logbooks (printed on water-resistant paper!) to chart daily abstractions, are being distributed to water-point committees. DRWS will feed all data gathered into a central database to monitor consumption and abstraction. Booklets are being distributed with guidelines on how to set up and run water committees. Participation by women varies, but they are generally underrepresented in water structures, despite their contribution to the development of schemes.

During South Africa's war with Angola and, at the time of the liberation movements in Namibia, water was supplied to rural communities at no charge in an attempt to win the hearts and minds of local residents. The provision of free water by the central government entrenched the idea that water is a free and abundant resource provided by government. Attempts to change this perception are now under way. Given low affordability levels in many areas, the government is not aiming at full cost recovery. However, payment may lessen wastage. Water tariffs are being introduced throughout the country, and with modest regular increases, full cost recovery on rudimentary schemes might be achieved by 2007.

Urban water supply

Most centres with populations in excess of 2 000 are supplied from a state water scheme, managed by DWA, which draws water from ground and surface resources. DWA sells bulk water to local authorities, where these exist.

Urban water supplies are likely to be placed under the most stress because of rapidly increasing population densities, higher per capita consumption levels, remoteness of most towns from perennial rivers, and high evaporation rates in urban water-supply dams.

At current rates of annual increase, Windhoek could start to run short of water by 1998. Water tariffs were recently raised by 30%, and water consumption exceeding 60 m^3/month per household or enterprise is billed at 5.30 ZAR/m^3. In response, average consumption has dropped by 25%, although it is unclear whether this reduction is in response to price cuts or moral obligation. Anticipated demand increases are such that, within the next 10–15 years, Windhoek will probably need access to the nearest perennial water source, the Okavango River, 800 km away. The final stage of the Eastern National Water Carrier would then have to be constructed to convey the water.

Construction of the Okavango link could also be postponed if groundwater resources north of Tsumeb prove to be as abundant as are currently anticipated. However, the longer Namibia delays abstraction from the Okavango, the more likely it is there will be competition for this water source from other users, as well as objections from interest groups.

Inadequate maintenance of water mains and distribution schemes is a major problem in Namibia and leads to significant wastage. Again, tariff increases may resolve this to some extent. Infrastructure maintenance is expected to improve after commercialization in late 1995 of the bulk water-supply section of DWA. This will also improve management, cost-effectiveness, and planning flexibility.

As a commercial utility operating a capital-intensive system, DWA will probably seek to maximize the sale of water, even though such sales could undermine long-term strategies to conserve water. By the same token, maximizing revenue by increasing tariffs would curb demand increases, though by how much is not clear.

Role of NGOs

Relations between the NGO sector and government are generally very good. Representatives meet monthly in water and sanitation forums in both Windhoek and Cuvelai to discuss priorities and coordinate development. Because DRWS has extremely few people on the ground, it is imperative that it maintain good relations with the NGO sector, whose role at present is crucial in setting up and maintaining local water schemes.

Major constraints and recommended research

Population and environment

PROBLEM — The two key resource issues facing Namibia are population growth and environmental degradation. Complementing these are a range of subsidiary issues, such as overstocking, denudation, erosion, and desertification. Stock-reduction schemes are a major issue.

RESPONSE — Public education campaigns are being developed by the government.

RESEARCH — Community-based socioeconomic incentive systems that match livestock levels to the carrying capacity of the land need are a priority.

Technical skills

PROBLEM — Government capacity to implement its policies is limited by staff constraints. Namibians with the necessary technical skills are leaving the public sector for the private sector, where salaries are up to 50% higher. Some of these posts have been filled with expatriates on contract, who are supposed to train local personnel, but few local personnel are available in the public sector to be trained. The result is a growing dependence on expatriate technical personnel on short-term contracts and an increasing use of private consultants. Commercialization of the bulk-water sector will exacerbate the technical-skills shortage in the public service.

RESPONSE — None.

Local skills

PROBLEM — The rural water-supply sector will require time for training people for the posts currently being created, but relatively few posts call for high technical training.

RESPONSE — A major recruitment and training program is under way.

Urbanization

PROBLEM — Namibia's urban areas are growing at a rate of 5–11% per year. Water consumption in urban centres is far higher than that in rural areas because of individual household connections. This poses a particular problem because few Namibian towns have sustainable local water sources.

RESPONSE — Raised water tariffs were introduced in 1995, and alternative supply sources are being explored and developed.

RESEARCH — The demand–price elasticity characteristics of urban water consumers need to be better understood so that tariff policies that control demand increases more effectively can be designed.

Cost recovery

PROBLEM — Urban water tariffs do not achieve full cost recovery and barely cover operation and maintenance (O&M) costs. The result is a decaying water infrastructure that leaks and wastes water. Water tariffs need to be trebled to cover the full cost of current delivery.

RESPONSE — The government sanctioned a 30% tariff hike shortly after the first postindependence election, in December 1994.

Bulk infrastructure provision

PROBLEM — Management and development of bulk water supply are underfunded and constrained by bureaucracy.

RESPONSE — Privatization of bulk water supply was scheduled for late 1995.

Complacency over water availability

PROBLEM — The government's ability to provide high-technology solutions to many of Namibia's water problems has lulled most residents into believing that the water shortage can be overcome. Urgent public-education campaigns are necessary to promote awareness of the need for water conservation and better resource management.

RESPONSE — Public-education campaigns have been designed and will be launched soon.

RESEARCH — To prioritize user sectors and to improve the targeting of drought restrictions and scarcity-education campaigns, a better understanding is required of the intrinsic value that the various user sectors attach to water.

Rural water-supply infrastructure

PROBLEM — Water-supply equipment in rural areas is too heavy and complex for its primary users (rural women) to maintain. Water committees are unlikely to be able to develop the technical and managerial skills that DRWS requires of them.

RESPONSE — A very ambitious plan to train a corps of local water minders, assisted by local rural water extension officers, is being developed.

Planning

PROBLEM — No detailed master plan exists, and there is little coordination among government departments. Furthermore, no government department is taking responsibility for coordinating and implementing policy around sanitation.

RESPONSE — A Water and Sanitation Committee, representing all stakeholders, has been suggested. Its role would be to improve coordination and advise cabinet.

Legislation

PROBLEM — Namibia's *Water Act* is based on South Africa's *Water Act* and is clearly inappropriate to Namibia's needs.

RESPONSE — A draft revision exists but has not been finalized.

South Africa

Water supply and demand

Water availability

The greater part of South Africa is semi-arid and subject to variable rainfall, droughts, floods, and high evaporation. The mean annual rainfall is only 500 mm, which is 60% of the world average. In addition, this rainfall is poorly distributed relative to areas experiencing economic growth. Only a comparatively narrow region along the eastern and southern coastline is moderately well watered, whereas the greater part of the interior is arid or semi-arid. Given that 65% of the country receives less than 500 mm of rainfall annually (the level regarded as the minimum for successful dryland farming) and 21% receives less than 200 mm, South Africa's existing and future development depends to a large extent on the state's ability to move water in bulk from the well-watered regions to the centres of settlement and industry in the drier regions.

Water supply

Under the apartheid regime, the Department of Water Affairs and Forestry (DWAF) practiced the art of large-scale interbasin transfer, to the acclaim of the international water industry. However, DWAF performed this role exclusively on behalf of white South Africa and those nonwhite population groups that were allowed to reside outside the Bantustans. The supply of water in the so-called independent and self-governing homelands was the responsibility of the individual Bantustan administrations, which undertook the task with varying degrees of success.

Today, an estimated 16 million people in South Africa, 40% of the population, do not have adequate supplies of safe drinking water. This uneven situation is the result of a number of factors:

- the poor performance of Bantustan administrations, especially in maintaining existing infrastructure;
- limited state development capital;

- rapid urban settlement, which is outpacing the development of new water-supply infrastructure;
- inappropriate water allocations to commercial agriculture, often at the expense of primary users; and
- poor control over the abstraction and pollution of water resources.

To meet minimum needs in rural areas, an extra 120×10^6 m^3 of potable water must be made available each year. Relative to total demand, this is a small amount, but a significant portion of this new demand will have to be supplied in semi-arid areas, where very little surface water is available, infrastructure is poor, and population density is low.

Nonconventional sources of water

South Africa's relative abundance of first-world technology and skills has led to the investigation and development of several nonconventional ways to augment water supplies. None of these can be said to have been motivated by actual water stress. The achievement of first-world water-supply and treatment standards seems to have been the main driving force. The two most notable areas of research and development were in rainfall stimulation (cloud seeding) and ultrafiltration technology (desalination).

CLOUD SEEDING — Cloud seeding was practiced in the Bethlehem area of the southern Orange Free State, where it was found to benefit the yield of farm dams but not the runoff from the Vaal catchment. The program has since been moved to the escarpment area of the eastern Cape, where some measure of success is being experienced in increasing the rainfall over commercial tree plantations. This has been an expensive project, and there have been recent investigations to determine whether the money could have been better spent on improving water conservation.

DESALINATION — Ultrafiltration technology was largely developed in South Africa to deal with the wide range of industrial and mining pollutants. By law these have to be returned to the channel of origin, but there is usually insufficient dilution potential there to render the effluent harmless. Although there is great potential for augmenting South Africa's water supply by desalinating seawater and brackish groundwater, the costs are still prohibitive. As such, this state-of-the-art technology has not yet been applied to domestic water supply, except in exceptional circumstances.

Water demand

Out of 22 main drainage basins in South Africa, six are already experiencing deficits anywhere from 1×10^6 to 122×10^6 m^3/year. depending on the severity of localized drought conditions. Another six have surpluses ranging from more than 1×10^9 to 4×10^9 m^3/year. At the projected rate of water-demand increase, several basins will experience deficits of more than 1 000 m^3/year by 2010, but a larger number will still continue to have healthy surpluses of that same volume.

In April 1994, the new Government of National Unity, led by the African National Congress, came to power with a clear set of policy objectives to address the gross distortions of the apartheid era. The Reconstruction and Development Programme (RDP), for example, devotes special attention to rural and urban water supply and outlines a number of specific targets, with clear deadlines. Within the short term, generally understood as being by 1997 (although this now seems unlikely), the new government aims to provide all rural households with a clean, safe water supply of 20–30 L/person per day within 200 m and an adequate sanitation facility for each household. By roughly 2002, it aims to achieve an on-site water supply of at least 50 L/person per day. It is imperative that these objectives are met as soon as possible, yet capacity and logistic and financial constraints suggest that meeting them on schedule will be a major challenge. It should be noted that, within the first 6 months of office, the new government announced plans to construct projects that will improve the supply of water to 1.2 million people, many of whom are rural dwellers. Existing and projected sectoral water demand for South Africa is shown in Table 3.

The bulk of the nation's available water resources is assigned to commercial irrigation. Although the land-reform debate is starting to acquire substance in South Africa, many rural people are realizing that water is the primary limiting factor governing the allocation of land to emerging small-scale farmers. Because existing rural water supplies are largely fully utilized, water for rural domestic use and new agricultural development will have to be reallocated from large-scale commercial farming operations. This will entail some combination of the expropriation of water rights, which could prove costly and controversial; the development of more storage, which is expensive and not always possible; and some sort of differential water-pricing strategy that will force commercial agriculture to improve efficiency and stop irrigating low-value crops.

WATER-DEMAND MANAGEMENT — The commercial irrigators of South Africa are not the only water users that should be considered for cost-effectiveness or efficiency improvements. Indeed, by international standards, these irrigators are

Table 3. Current and projected demand in South Africa by sector.

	Demand (× 10^6 m^3)	
Sector	1993	2010
Domestic	1 516	3 000
Industry	1 031	2 500
Municipal use	90	200
Urban losses	280	500
Power generation	224	400[a]
Mining	466	600
Irrigation	8 254	11 500
Stock watering	264	350
Forestry	1 284	1 700
Nature conservation	2 994	5 000

[a] Losses to evaporation in cooling systems, plus water used in ash-disposal systems.

among the more efficient in the world. The trend of the last 12 years toward drip and microirrigation systems, which was prompted by the shortage of water rather than by price, has resulted in leaching fractions of less than 15% in many areas. However, bulk delivery of water to irrigators still incurs losses in excess of 30%.

The other major user sector that needs to consider its consumption levels is the urban domestic one. Recent surveys conducted in South Africa's middle-class suburbs revealed that the conservation-threshold price of water (that is, the price at which householders would implement conservation measures) could not be determined with any accuracy because it was too much in excess of what was currently being paid for respondents to identify a specific price. This suggests that there may be considerable scope for raising revenue in South Africa by means of levies on urban water sales.

Clearly, there is the potential to curb South Africa's thirst for more water by introducing demand-management strategies. Furthermore, there is also some urgency to do this, given the new government's intention to expand the economy at a rate of around 5% per year. Unfortunately, there is little indication that policymakers are thinking along these lines.

Water-management systems

National overview

Responsibility for water supply in South Africa is divided among central, regional, and local authorities and nonprofit bulk-supply authorities (water boards), with the central government's DWAF managing the overall policy framework. DWAF is responsible for operating many of the country's major dams, setting policy, issuing forestry permits (based on estimated water use and runoff reduction), and coordinating long-term water-resource development. In the past, DWAF worked closely with provincial and ethnic homeland governments but had no jurisdiction over the four nominally independent homelands.

The former homeland governments are currently being reabsorbed into central and regional administrations. Nine new provincial governments have been established in place of the former provincial and homeland administrations. Despite the detailed wording of the new constitution, there is still uncertainty concerning the division of responsibility between central and regional governments, especially because the new regional governments are keen to assert their independence of central government. Water management is a responsibility of the central government, whereas provincial and local services are the responsibility of regional governments.

The anticipated lack of capacity at the provincial level, coupled with the rigid service-level targets of the RDP, has forced the central government to consider the establishment of water utilities (that is, nonprofit, democratically controlled water boards) to implement water-supply and sanitation works at the subprovincial level. Although these new utilities have still to be formed, it is widely recognized that the needs of rural communities will only be met with a collective and concerted effort from government, parastatals, NGOs, the private sector, and the communities themselves. Moreover, to benefit from the RDP, the communities must demonstrate both consensus and determination regarding these issues. The backlog in water-supply and sanitation services, the rapid urbanization rate, and rural population-growth rates of more than 3% per year present too great a problem for any single agency to assume total responsibility.

Rural water supply

Because of the policy of separate development, rural water supply in white South Africa, as administered by DWAF, focused on water supply for commercial agriculture, whereas water supply for the inhabitants of the ethnic Bantustans was allocated to homeland governments. In general, the latter lacked the resources, capacity, and motivation to introduce and maintain sustainable water-delivery systems, although there were isolated success stories.

More than half of all rural people rely on unimproved water sources: streams, rivers, and unprotected springs. This direct dependence on natural water sources has made many communities highly vulnerable to droughts, to increases in water-abstraction patterns, to upstream land-use changes, and to effluent discharges. Current water conflicts and detailed catchment analyses are starting to indicate that rural water resources may have been overallocated, going beyond the safe yield of catchments and dams. Furthermore, nearly all of South Africa's surface water is unsuitable for human consumption in an untreated state, largely because of contamination by human, animal, and industrial waste. This water must be treated if any reduction in rural waterborne diseases is to be achieved.

In response to this problem, DWAF is setting up the Directorate of Rural Water Supply and Sanitation to support rural communities. The new directorate will strive to facilitate, rather than implement, water management. NGOs, local authorities, water boards, the private sector, and regional authorities will take primary responsibility for implementation. Local water committees will coordinate the rural water supply. It is anticipated that the current mapping of groundwater potential throughout the country will assist in satisfying rural water demand.

Urban water supply

South Africa's urban areas were historically segregated into white, Asian, coloured, and black areas. Water supply in white areas is generally excellent. Because African urbanization was forcibly discouraged, black townships outside the homelands were not designed to accommodate large populations. Since the lifting of influx controls in 1986 there has been a dramatic shift away from covert settlement in overcrowded township houses and back yards toward shack settlement on the urban periphery. Existing township infrastructure is proving inadequate to meet the demands now being placed on it. Politically motivated rent and service boycotts have deprived many local authorities of the revenue needed for maintaining and upgrading systems, and maintenance capacity has been deteriorating steadily for some time. Nearly 50% of the water supplied to Soweto is lost through leakage. The new political dispensation does not seem to have defused the crisis in many local authorities as residents grow impatient at the state's failure to deliver widespread civic improvements. Local government elections held in November 1995 give more legitimacy to the local councils that levy tariffs, but the perception that urban services can be had for free is now deeply entrenched in many black townships.

In the last 5 years, dense, informal settlements have sprung up around the periphery of towns and cities, such as Johannesburg, Durban, and Cape Town.

Growth is extremely rapid. In the space of a month, shack settlements of several thousand people can develop on vacant land, and local authorities are battling to provide even rudimentary services. Most residents rely on public standpipes in adjacent settlements or on water tankers. Sanitation is often nonexistent, and already there have been several outbreaks of typhoid and other waterborne diseases.

The involvement of the residents of informal settlements in water management or local administration was initially discouraged by the last government for fear of bestowing some form of legitimacy on such settlement practices, as they are perceived to be an illegal activity. The new government has adopted the policy of encouraging dialogue with informal community leaders and encouraging the provincial governments to give priority to their needs.

Cost recovery

The majority of rural users regard water as a free good that the government must provide in abundance. Government makes little attempt to recover the real costs of rural water supply from the beneficiaries, and rural consumers are required to pay a nominal tariff that is seldom consumption related, and then only if the water provided is close by. Very few people actually pay even this nominal tariff, and government agencies lack the capacity to enforce payment. In urban areas, mounting debt from long-standing rent and service boycotts in many areas is being covered by government through interdepartmental cash transfers, which depletes funds for new services and housing.

Payment for sanitation in rural areas has generally entailed the repayment of loans on a small scale for voluntary, owner-built pit latrines. However, disposable income is extremely limited in the impoverished rural areas, and other priority needs, such as water supply, food purchases, health, and education, are often considered more important. Service-payment schemes for RDP sanitation systems have yet to be developed and implemented.

The nonrecovery of even the O&M costs of water-supply and sanitation services has all the ramifications that have been experienced elsewhere in the world, such as

- failure by the community to identify with the scheme so long as it is state owned;
- lack of respect for the scheme, leading to abuse, theft, and vandalism, thereby raising maintenance costs;
- inappropriate system design resulting from lack of effort to reconcile the needs of the community with its levels of affordability and its capacity for managing sophisticated systems;

- increasing national O&M cost burden, which reduces available funds for new schemes and leads to a preoccupation within government with cost cutting; and
- wastage of water.

Homeland governments lacked the legitimacy and political will either to introduce or to enforce better cost recovery. Often, zero-cost-recovery water-supply schemes were seen as a way to foster rural political support, something that took on greater importance in the run-up to the 1994 elections. Today, in some large-scale rural reticulation schemes, the inherited policy of zero cost recovery completely undermines the reliability of the service, as people are not motivated to turn off yard taps and prefer, instead, to run the water continuously on vegetable gardens and lawns. This practice leads to pipeline-pressure reductions, with the result that many villages farther down the line do not receive water and have to rely on tanker deliveries.

The task of achieving a partial cost recovery, sufficient to meet at least O&M costs, now falls on the new government. The policy on this has been advanced in RDP documentation, where it was indicated that service charges would have to be levied on RDP infrastructure projects. However, the implementation of this policy in the existing environment, where service delivery has been highly politicized and welfare expectations run high, has still to take place.

DWAF plans to entrench the principle that all water consumed has a price. A low-level "lifeline" tariff has been suggested for low-income consumers, and higher tariffs on a sliding scale for other consumers could help finance a more equitable and sustainable water supply to all. Higher water tariffs for irrigators are also possible.

On payment, the achievements of a number of independently minded rural communities deserve recognition. Either as a result of objections to incorporation in homeland states or because of disfavour with homeland politicians, a number of rural communities in South Africa were left to fend for themselves during the 1980s and early 1990s. To survive, they established the foundations of an interim local authority, often with strong inputs from women. This authority collected funds within the community and approached NGOs for assistance with basic-service provision. In some instances, this new-found capacity has led to the development of rural industries and enterprises. The interesting feature of the many water schemes originated by the community is the reportedly high level of assurance of supply. This assurance is due essentially to an absence of government intervention in the development of the schemes. In such cases, it is not uncommon to find the local vehicle mechanic maintaining the village water pump.

Catchment management

Historically, natural-resource management has tended to focus on "separatist conservation," rather than viewing humans as an integral part of the natural environment. Hence, no catchment-management plans that can be implemented exist beyond statements of general intent. For example, dense settlements, overgrazing, and overburning have led to severe erosion, particularly in the former homeland areas. Diminished infiltration is affecting local groundwater supplies, flooding is increasing in frequency and severity, and the silting of rivers and estuaries is widespread. The impact of this degradation not only affects people in the immediate vicinity (the rural areas), but also will, in a short space of time, affect water supplies to the main metropolitan areas.

Broad policy statements are difficult to implement on the ground, and neither the Department of Agriculture nor DWAF has the power or resources to ensure better resource management in the important water-supply catchments. In the former homelands, environmental management was largely ignored. Instead, the focus tended to be on tourism-related conservation, rather than on veld rehabilitation and community education. Effective public-awareness campaigns are an urgent priority. However, in the absence of feasible management policies and the necessary development to alleviate rural poverty, progress is likely to be limited.

Major constraints and recommended research

Top-down development practices

PROBLEM — The historical approach by government agencies has been top down and paternalistic, with the emphasis on technical rather than institutional development. *Community participation* tended to mean "consulting the chief or headman about the siting of a borehole." Communication was generally one way and addressed exclusively to men.

RESPONSE — The new minister of DWAF has committed the department to the principle of bottom-up development, involving consultation with all stakeholders, particularly women. Extensive staff recruitment and management reeducation programs will be necessary to achieve this, as DWAF has tended to be white, male, and technocratic. Policies are being put in place to address this.

Poorly developed capacity in rural areas

PROBLEM — The existing rural water-supply system has fostered dependence and stifled initiative, without being able to meet community expectations. Organizational capacity at the local level is generally poorly developed, and if local committees are to play the role outlined for them by the new government, considerable training will be required in organizational development, basic administration, bookkeeping, and rudimentary maintenance skills.

RESPONSE — In conjunction with NGOs, the government is formulating strategies to address training and organizational development. Also, in response to the realization that well-developed technical skills often exist in rural communities and merely require reorientation toward water-system O&M, there are moves to help establish small-scale water and sanitation entrepreneurs in rural areas.

Pollution of water-supply aquifers

PROBLEM — Contamination of water sources and aquifers has become a very serious problem, in part because of the rapid growth of dense settlements with poor sanitation infrastructure. Considerable contamination of groundwater has already occurred in many places, and it will be some time before government and community agencies have the resources to introduce better sanitation techniques and alternative water supplies.

RESPONSE — Improved sanitation is one of the RDP's priorities, and DWAF has assumed some measure of responsibility for ensuring improvements do occur.

Overstocking of grazing veld

PROBLEM — In the past, tribal chiefs were responsible for ensuring that lands under their jurisdiction were well managed. However, the erosion of their legitimacy in some parts of the country, coupled with incentives that encourage chiefs to maximize the number of households (with livestock) in the communities, has greatly reduced their enthusiasm to introduce new patterns of stock management. Government attempts to reduce stock numbers in tribal areas and forcibly introduce different stock-management regimes have often been politicized and have generally failed, although there have been documented successes in controlling the stocking rates of white farmers.

RESPONSE Politicians have shied away from tackling this problem, both because of the cultural importance of cattle to many African people and because it tends to raise questions about land distribution in South Africa. It is one of the most pressing problems facing sustainable water management.

RESEARCH — There is a need to develop community-based socioeconomic-incentive systems that will match livestock levels to the carrying capacity of the land. This may entail developing closer links between livestock farmers in tribal areas and the meat-products processing and marketing industry.

Water for the environment

PROBLEM — As a result of weak enforcement of environmental-conservation policies and a poorly coordinated environmental lobby, when water stress does occur, the natural environment invariably suffers. The water needs of wetlands, riverine habitats, and even conservation areas, such as Kruger National Park, have generally been overlooked.

RESPONSE — The Water Research Commission, in conjunction with the Foundation for Research Development, embarked on a program to determine the water needs of the natural environment in the mid-1980s. It remains to be seen whether the state will adopt the recommendations of this research, in view of the stiff competition for water from rural communities, although it is likely that the state will accept the concept of the aquatic environment being the resource from which water can be drawn, up to a certain point. In other words, the natural environment would no longer be viewed as a user but as the source of water and thus would be entitled to a reserve beyond which further abstractions would not be allowed.

RESEARCH — The contribution of freshwater systems to the economic performance of various sectors, such as agriculture and tourism, needs to be better quantified.

Legislation

PROBLEM — Responsibility for implementing the *Water Act* is dispersed amongst a myriad of authorities operating at various levels of government within South Africa and the independent and self-governing states. Moreover, South African water law is derived from European law and presupposes an abundant supply of water. Thus, the emphasis is on allocation, rather than on integrated scarce-resource management.

RESPONSE — Under the new government, DWAF is drafting legislation to consolidate and rationalize water legislation into one uniform body of law. From there, the nation's water law itself will be revised to meet current conditions.

Drought management

PROBLEM — Droughts are still seen as exceptional, rather than inevitable and predictable. Drought-relief schemes during the 1992–93 period raised awareness among government officials that, in most places, the major problem was not that assured water was absent but that existing infrastructure had broken down, which forced communities to revert to traditional sources, which were soon exhausted.

RESPONSE — Better awareness of the need for ongoing maintenance programs involving local users exists, but few initiatives are under way.

Coordinating the activities of NGOs

PROBLEM — NGOs have played a pivotal role in installing improved water supplies in areas inadequately addressed by government. However, the rapid proliferation of NGOs in the late 1980s and early 1990s resulted in poor coordination and communication between organizations.

RESPONSE — A new NGO, the Mvula Trust, has been established to fund rural water and sanitation schemes and to coordinate their funding and implementation.

Swaziland

Water supply and demand

Water availability

The water resources of Swaziland may be described in terms of the four main geographical areas, each stretching from north to south. From west to east, they may be described as follows:

- Highveld has rainfall of 1 000 – 1 200 mm/year and abundant surface and groundwater.
- Middleveld has rainfall of 600–800 mm/year and good groundwater, but its surface water is unreliable.
- Lowveld has rainfall of less than 600 mm/year, little surface water, and few successful boreholes.
- Lebombo Plateau is relatively wet in the northeast, which has rainfall of 650 mm/year. Most rivers in the northeast have small irrigation dams for sugar, citrus, and other crops. The southeast is in a rain shadow and receives rainfall of only 400–500 mm/year. Domestic supply relies on boreholes and a little surface water.

Water supply

The traditional sources of water are springs and rivers, which are shared with livestock. The three main river systems affecting Swaziland are the Komati River, the Usutu River, and the Ngwempisi River, all of which flow east from South Africa through Swaziland toward Mozambique.

The Komati River hosts a number of impoundments on the upstream South African side. These are used for supply of cooling water for coal-fired power stations and water for irrigation. Commercial timber plantations in both South Africa and Swaziland further reduce the runoff from this catchment. Historical agreements with South Africa have allocated a portion of the Komati River flow to Swaziland. This allocation has generally been more than could be used by Swaziland. However, the droughts of the 1980s and 1990s, coupled with increased irrigation abstractions upstream of Swaziland, have greatly reduced the flow in the Komati River as it returns to South Africa. This has partly been the motivating force behind the construction of the Driekoppies Dam on the Lomati River (a main tributary of the Komati River), on Swaziland's eastern border.

A second dam in the Komati catchment, at Maguga, has been proposed. If it goes ahead, it is to be funded 40% by Swaziland and 60% by South Africa. However, Swaziland has been slow to initiate the processes necessary to keep project negotiations on schedule. Maguga will have virtually no impact on domestic water consumption, as its primary purpose will be hydroelectric-power generation and irrigation, especially of sugar, citrus, and other crops.

Water demand

Swaziland's current and projected water demands are shown in Table 4. Official estimates are set at 1.2×10^9 m^3/year for irrigation, based on the register of permits allocated for irrigation. However, actual use is estimated to be far less, at around 0.4×10^9 m^3/year; no precise figures exist.

Table 4. Current and projected demand in Swaziland by sector.

	Demand ($\times 10^6$ m^3)	
Sector	1994	2016
Urban domestic	5.77	16.35
Rural domestic (estimated)	4.60	10.00
Industrial	5.71	13.94
Irrigation	400.00[a]	500.00
Forestry	120.00	130.00

[a] Official estimates are set at 1.2×10^9 m^3/year for irrigation, based on the register of permits allocated for irrigation. However, no precise figures exist for actual utilization, which is estimated to be far less, at around 400×10^6 m^3/year.

Nearly 30% of the population of Swaziland lives in urban areas, and this proportion is increasing rapidly as people leave the rural areas in search of work. Political change in South Africa has led to a measure of disinvestment from Swaziland in preference to its more developed neighbour. This is expected to place greater pressure on the commercial agricultural sector to generate jobs and wealth, which in turn will probably increase the demand for irrigation water.

Water-management systems

National overview

Swaziland is a constitutional monarchy, and the royal palace strongly influences decision-making. There is no single institution outside the monarchy with the power to coordinate water policy in Swaziland. Authority is dispersed among several government departments, each of which seems eager to cede responsibility. Despite much discussion and an agreement in principle taken 6 years ago, the proposed National Water Authority, with the powers to gather information, formulate policy, plan development, and oversee implementation, has still not been set up. Development planning falls primarily under the Ministry of Economic Planning, whose priorities do not necessarily address resource management and sustainable water delivery.

There is no comprehensive national water-development strategy or master plan in Swaziland, and rhetorical commitments to water-resource development have not been matched by the necessary financial and human-resource commitments. For example, until mid-1995, the Rural Water Supply Board depended entirely on external donors for its existence. The Government of Swaziland has now made a formal commitment to partially fund the Rural Water Supply Board (now renamed the Rural Water Supply Branch [RWSB]). Although this limited government funding will lessen uncertainty about the future of RWSB, there are no guarantees about the size of the annual budgetary allocation from government from year to year. Consequently, RWSB will remain dependent on the donor community for much of its activities.

Part of the reason for this complacency is that Swaziland is well provided with surface water. However, much of this water is unsafe for human consumption, largely because of human and animal fecal contamination, and the failure to invest in sustainable rural water infrastructure is reflected in high infant-mortality rates, widespread diarrhea, and a range of other waterborne illnesses.

Estimates of safe water-supply coverage vary widely. Between 20 and 40% of the urban and peri-urban population and between 45 and 55% of the rural population do not have access to potable water. Given that 70% of people live in rural areas, this imbalance in water access reflects a bias toward higher quality service provision in urban centres.

Swaziland has three main water institutions, described as follows:

- The Water Department, within the Ministry of Natural Resources and the Environment, has few powers and resources. Its main activity is managing water for irrigation.
- RWSB falls under the Ministry of Natural Resources and the Environment but relies heavily on external funding. It was set up with donor funding and NGO support during the United Nations Decade of Water and Sanitation.
- The Water Services Corporation was privatized in August 1994 to facilitate better planning, budgeting, and overall management of urban water supplies. It remains answerable to the Ministry of Housing and Urban Development.

The Rural Water Supply Board (now Branch)

For many years the Government of Swaziland regarded RWSB as a temporary parastatal institution. Short-term and uncertain funding severely undermined its effectiveness. Most of its staff were in temporary positions and received little training, and training in community development was largely neglected. In mid-1995, the Rural Water Supply Board was formally reconstituted as a department of government and renamed as a branch (as mentioned earlier). It is too soon to tell what impact this change will have on the internal workings of the branch.

RWSB emphasizes low-cost, community-initiated water projects. People wanting an improved water supply are required to form a water committee and to collect contributions for O&M. When RWSB is satisfied the committee has shown sufficient commitment, it applies for donor funding. Funding, however, can take several years to secure. RWSB technicians install gravity-fed systems from reliable springs, wherever feasible, and sink boreholes where necessary. Where the depth is too great for handpumps, RWSB installs electric pumps wherever possible. Diesel pumps are rarely installed now because of theft of the pump and fuel. The use of electricity has raised the cost, and affordability is a major problem for many households.

A major factor complicating rural water provision is the entrenched cultural preference for scattered homesteads, rather than close rural settlements and villages. Among other factors, this raises the cost and difficulty of rural water supply. Once a water scheme is installed, many water committees lapse. Of those that remain, only half maintain an ongoing water fund. People in rural areas are supposed to contribute a monthly flat rate, usually 6 or 7 ZAR, for O&M to the local water committee, but payment and collection rates vary. In theory, nonpayers are not allowed to take water, but in practice such water theft is hard to police. Vandalism by those excluded and the health costs of reverting to traditional water sources are far more expensive. Maintenance capacity is, therefore, limited. Although 45% of rural villages are serviced by improved water schemes, not all of these schemes are in working order, because of problems with communication, transport, and the shortage of technical staff. Many rural water-supply schemes are also too complex to allow for greater community involvement. Moreover, as water minders are unpaid and usually untrained, very few settlements have a permanent water minder.

The Urban Water Services Corporation

The Urban Water Services Corporation (WSC) is understaffed and depends on expatriate technical advisors. Recent internal restructuring has distracted WSC from the need to make urgent decisions about infrastructure upgrading and expansion. At current levels of consumption and urban growth, Mbabane will start running out of water in 1998. WSC has not decided yet been how to address this problem.

Current urban-development strategies stress full cost recovery but do not achieve this, as tariffs only cover O&M costs and do not provide for expansion and development. Formal settlement areas, on the other hand, have individual, metered connections, and revenue collection is well administered. These high-level, metered reticulation networks in formal settlements coexist uneasily with a dearth of infrastructure in the informal settlements. An estimated 140 000 people in urban and peri-urban informal settlements are without running water, and settlements in the greater Manzini area are growing at a rate of 5% per year. Urban community structures, however, are not involved in water management.

A few public standpipes were installed by WSC in informal settlements in the 1980s, after a cholera outbreak, but no attempt was made at cost recovery. Several pilot projects are now under way to get standpipe users to pay for water, despite the lack of precedent for this in Swaziland. One expedient is lockable standpipes, with keys for those who pay a flat rate. However, there are problems with vandalism by those locked out. Water kiosks seem to be more successful.

The NGO sector

Coordination between the government and some elements of the NGO sector is poor, leading to inefficiencies and duplication. RWSB is highly critical of some

NGOs, which, it argues, often install inadequate water schemes fitted with nonstandard equipment that RWSB is then obliged to maintain.

International institutions

As a result of the construction of the Driekoppies Dam, an international water-management institution has been established through an agreement between Swaziland and South Africa. The Komati Basin Water Authority will have the task of monitoring land use and runoff within the Komati catchment, which includes parts of Swaziland.

Major constraints and recommended research

Drought relief

PROBLEM — As Swaziland has only one major storage dam, Mnjeli, used mainly for irrigation, and has inadequate water-supply coverage in rural areas, the country is extremely vulnerable to drought. An elaborate scheme to erect water tanks in stressed areas and have them filled by government tankers failed conspicuously when it was realized that no funds had been voted for the O&M of the tankers. Moreover, the logistics of supply in the rural areas proved prohibitive. Local residents expressed their frustration in some areas by vandalizing the empty tanks.

RESPONSE — A borehole-drilling program provided by RWSB on behalf of the Disaster Relief Task Force is under way to improve rural water supply, particularly in vulnerable areas.

Natural-resource management

PROBLEM — Poor resource management in communal lands, controlled by the chiefs on behalf of the King, is compounding erosion and aggravating the sedimentation of rivers and reservoirs in the southwest. The main water-supply catchment for the greater Manzini area (the industrial hub of the country, with rapidly growing informal settlements) lies in badly degraded communal lands. The combination of steep slopes, erodible granitic soils, overgrazing, and high-intensity rainfall has led to major sedimentation. The silting of dams and reservoirs is, therefore, a serious problem, particularly in the Matsapha–Manzini area.

RESPONSE — Major dredging operations have been necessary for the past year to improve the storage capacity of the Matsapha Dam, but the soil-conservation measures of the Department of Agriculture have proven ineffectual.

RESEARCH — Community-based socioeconomic-incentive systems that encourage the matching of livestock levels to the carrying capacity of the land need to be developed.

Institutional coordination

PROBLEM — There is little overall coordination among agencies implementing water and sanitation programs. Communication between the various agencies happens largely at a personal level, rather than an institutional level. There is no monitoring system at the national level.

RESPONSE — There has been no apparent action.

International water sharing

PROBLEM — South Africa abstracts heavily from two of the three main rivers entering Swaziland, with six dams on these rivers. Swaziland wants this water for irrigation, but government officials maintain that flows through Swaziland are declining because of South Africa's dams. Bilateral negotiating mechanisms were introduced in 1979, after five of the South African dams were completed.

RESPONSE — Because of Swaziland's size and location, government officials in Swaziland feel relatively impotent in asserting the nation's right to a more equitable share of river flows. Response has been correspondingly limited.

Water conservation measures

PROBLEM — In urban areas, the estimated volume of water lost between the water supplied and water billed is 60%.

RESPONSE — Private consultants were commissioned to investigate loss-reduction schemes, such as replacing valves, upgrading shutoff devices, and redesigning mains. However, government is stalling on the implementation of the recommendations, and there is no evidence of water-conservation measures being applied in the other user sectors.

Conclusions

This regional overview highlights the different ways water is supplied and managed in southern Africa, but it is not an evaluation of those water-management systems currently in place. The focus is on water stress and how governments, institutions, and local communities perceive such stress and respond to it. However, as water stress is often the product of institutional or management failure, this report tends to overlook the successes and achievements of the countries reviewed. Such achievements have been considerable, and the necessary planning to meet future needs is often well developed. Unfortunately, the resources (skills and finance) to sustain these successes and implement new schemes seldom exist. In the highly stressed areas, this has led to water management based on survivalism, sometimes at the expense of the environment and economic development. Thus, there is a need for a dualistic approach to water supply in the region: (1) to consolidate existing systems and ensure their effectiveness and sustainability; and (2) to meet future demands for both domestic use and economic development.

Despite shared characteristics and somewhat similar geography, each country reviewed has a distinctive approach to policy and has specific strengths and deficiencies. A number of problems concerning water supply are common to all four countries:

- a looming water shortage;
- little popular awareness that regional water resources are finite, coupled with a widespread perception that government has the ability to provide abundant water;
- inappropriate tariff structures, poor cost recovery, and problems in getting users to pay for the water supplied;
- an emphasis on installation of water-supply systems, rather than on their maintenance;
- inadequate water-management education, training, and support for rural users;
- serious environmental-degradation problems, particularly relating to rural land management;
- poor coordination among water-management agencies; and
- inadequate attention to sanitation.

More positively, there is a wealth of water-management experience in the subregion. Now that South Africa has moved beyond apartheid, targeted regional-cooperation initiatives may become possible, with benefits to the subcontinent as a whole. Lesotho, for example, has many years of experience in developing low-cost, low-technology, community-driven rural water schemes. Swaziland has experience in using Afridev pumps, which many regard as the most accessible and manageable borehole technology for rural women. South Africa has excellent technical expertise and an impressive record in the development of bulk water-supply infrastructure, and the model Namibia has recently developed for managing rural water supply has widespread application. However, there are some problems in the area of water management:

- lack of demand-management strategies;
- absence of any sustainable grazing-management systems in key water-supply catchments;
- poor record of skills and technology transfer, especially at local levels; and
- lack of government realization, policies, or actions regarding the need for water-supply education and training focused on rural women, even though women's community ties are strongest and their benefits from improved water-supply schemes would be the most significant.

Despite critical water shortages, either current ones or those expected in the not too distant future, policymakers pay little attention to curbing the demand for water or creating incentives for water conservation. The people of southern Africa believe that their respective governments can supply unlimited quantities of water indefinitely, and the governments themselves seem to be labouring under the same misconception. The attention being given to new and elaborate water-supply engineering projects by recently democratically elected governments suggests a fixation with supply-based solutions. Whether this is the preference of the politicians or of the engineers advising them is not clear. However, there does seem to be a reluctance of governments to control consumption or practices that threaten the sustainability of water supplies. Indeed, it has been suggested that some water-supply agencies are unsupportive of non-drought-related water-conservation initiatives because of the prospect of reduced revenue. Increased international encouragement for governments to consider and adopt one of the many types of demand-management strategies may well be justified. The development and tailoring of such strategies to the individual countries and situations could be a primary area of research.

Southern Africa is reasonably well off in terms of water-supply skills. However, these continue to be vested in a centralized minority, many of whom are expatriate. Apart from the recent training efforts of Namibia, which have still to bear fruit, there are few, if any, programs to transfer water-supply and management skills to individual rural communities. Furthermore, there is no evidence of governments' acknowledging the indigenous skills that have historically enabled rural communities to secure reliable water supplies before reductions in river flows resulting from upstream development and population increase. The belief that rural communities are simply incapable of looking after their own water-supply systems still prevails in many government departments, although it is seldom openly admitted. One result of this is that no attempt is made to educate communities about the reality of the water situation in the region and thereby provide a foundation for future water conservation.

Finally, broad consensus can be found in the development agencies of all four nations that there is considerable merit in focusing water-supply education and training on rural women. Their community ties are strongest, and they benefit most from improved water-supply schemes. However, this realization is not, as yet, reflected in either the policies or the actions of governments.

Strain, Water Demand, and Supply Directions in the Most Stressed Water Systems of Southern Africa Except South Africa and Namibia

Luke Onyekakeyah
Environmental Scientist and Consultant, Nairobi, Kenya

Introduction

A stressed water system is one in which degradation is taking place or where there is a threat to its capacity to continue providing adequate water supply, in quantity and quality, to households, communities, and nations.

This study of stressed water systems in the southern Africa subregion used three methods:

- Questionnaires were administered to identified expert contacts in all eight countries. Thirty-two questionnaires were administered, and 16 (50%) were returned.
- A 1-week visit to the two countries serving as study samples, Zambia and Zimbabwe, was made to obtain detailed data.
- Pertinent information in published materials was reviewed and analyzed to supplement the field data.

Respondents were requested to identify stressed local water systems. Although this system of identifying stressed water points had obvious shortcomings, it provided an acceptable source of preliminary data. Further scientific data could be obtained through appropriate research. Respondents noted that most of the regional water courses were under stress (Table 1).

Actual data on groundwater were generally lacking. According to Moyo et al. (1993), well-digging to extract groundwater began in the 19th century and penetrated to deeper and more effective sources with the availability of the diesel pump in the 20th century. There were no available statistics on the number of boreholes or well points, but it was estimated that over the last two decades the

Table 1. Stressed water systems in southern Africa.

Water system	Countries sharing water system	Estimated dependent population	Suspected source(s) of stress
Zambezi	Mozambique, Zambia, Zimbabwe	7 250 000	Industry, agriculture
Limpopo	Botswana, Mozambique, Zimbabwe	5 485 257	Industry, HEP
Save	Mozambique, Zimbabwe	5 250 000	Agriculture, HEP
Kafue	Zambia	2 000 000	Industry, agriculture
Okavango	Botswana	235 257	Drought, agriculture
Luapula	Zambia	2 000 000	Agriculture, industry
Orange	Lesotho, South Africa	1 625 000	Agriculture, HEP
Kariba	Zambia, Zimbabwe	4 250 000	Industry, agriculture
Tanganyika	Tanzania, Zambia	6 250 000	Agriculture

Note: HEP, hydroelectric power.

the number has increased tremendously, from 500 to 10 000. These are suspected to be depleting the groundwater reserves, the total quantity of which is yet unknown.

Against this background, I designed the survey to obtain factual information on the water-stress issue so that I could determine the most appropriate research agenda to address the region's strained water resources.

Sources of stress and alleviation strategies

A stress factor is any element or situation that obstructs the accessibility or quality of the water supply. Moreover, any factor or situation that puts too much demand on the available water supply, thereby making it unable to satisfy the projected demand, is regarded as a stress factor.

The incidence of stress also has a space–time dimension and can be classified as either temporal or permanent. Human activities, such as the discharge of industrial waste into the Kafue River in Zambia, limit the accessibility and quality of water, reducing its availability for different uses. As long as this situation persists, the population depending on the water source may be forced to explore alternative sources of water supply. However, this situation can change if the polluting industries adopt cleaner methods of production. The source of stress under these conditions has the possibility of changing over time.

On the other hand, stress can be permanent if its impact on the water-supply source cannot be reversed. For example, the construction of several hydro-power stations on the River Zambezi inevitably reduces the quantity of water available to the population living downstream. This situation may be irreversible, and the stress on the water system is permanent.

Stress elements that limit the quality of water for different uses can be

- purposive utilization, such as for domestic uses, agriculture, mining, and industry;
- unattended utilization, such as wastage and pollution; and
- purposive pollution, such as convenient dumping of waste before treatment.

Drought

Most of southern Africa lies in a drought-prone region, which experiences natural drought conditions caused by the prolonged absence of rainfall in both time and space. This problem affects many parts of sub-Saharan Africa. Within southern Africa, drought is considered a critical issue of development planning because it underlies many problems encountered in water-resources utilization. According to available records, drought conditions have occurred in roughly 20-year cycles: in the 1960s, in the 1980s, and again in 1991–92 (Moyo et al. 1993).

The occurrence of drought has serious consequences on water availability for domestic and agricultural use, reducing the water supply both in quantity and in quality. Reduced access to water particularly affects herding communities and forces the people and their livestock to concentrate and compete for water around water holes and other sources (Moyo et al. 1993). Drought, therefore, makes traditional farming and grazing lifestyles inadequate and unsustainable (Keating 1993) and results in poverty and starvation. An estimated 3 million people died in the mid-1980s because of drought in sub-Saharan Africa (Keating 1993).

Attempts have been made to deal with drought situations by increasing water supply. In Zimbabwe, for example, a major interbasin water transfer currently is being undertaken. The Zambezi Water Project is intended to pipe some of the water that thunders over the Victoria Falls through the arid landscape of western Zimbabwe, to help redress the effects of recurrent drought in that area

(ZERO 1993) and provide enough water for domestic and agricultural purposes. Similarly, in Zambia, the Community Management and Monitoring Unit (CMMU) of the Department of Water Affairs (DWA), under its Water, Sanitation, and Health Education (WASHE), is initiating a number of community-based field programs to educate the people in water conservation and to make water points easily accessible (Carty 1994). In addition, some nongovernmental organizations (NGOs) are working with communities to increase the water supply by constructing more boreholes (Connolly 1991). Similarly, the Zimbabwe Environment Research Organization has drawn up a regional research project to identify long-term strategies to cope with the recurrent drought in southern Africa, particularly in Zimbabwe (ZERO 1993). The project will cover three countries, Botswana, Mozambique and Zimbabwe, and has the following objectives:

- to document droughts within the three countries since 1900;
- to identify areas where drought impacts were most severe;
- to assess the region's current early-warning systems and analyze their strengths and weaknesses; and
- to investigate the impact of drought on vulnerable groups, especially women, children, the elderly, and the unemployed.

With the actions being undertaken at all levels, it is hoped that appropriate solutions will be found to deal with future drought situations in the subregion.

Soil erosion and sedimentation

Land degradation is a serious problem in catchments with large population concentrations and in high-rainfall areas. Misuse of land and vegetation by overgrazing and adoption of inappropriate agricultural practices has inevitably exposed large watersheds and made them susceptible to severe erosion. Arntzen and Veenendal (1986) showed that erosion is a serious problem in the region, exacerbated by periods of drought and intense rainfall. Several types of erosion occur. For example, sheetwash, rill erosion, and gullying have been identified in the hardveld of Botswana (Moyo et al. 1983). In Lesotho, a detailed study of soil erosion and reservoir sedimentation in two catchments of Roma Valley and the Maliele area revealed very disturbing levels of erosion and siltation (Chakela 1981). The rates of net erosion varied from about 100 to 2 000 t/km per year.

The incidence of soil erosion has severe impacts on water resources. Based on figures recorded at Little Caledon, Lesotho (Moyo et al. 1993), it has been found that soil erosion leads to the following:

- sediments in the floodplains of major streams and in reservoirs;

- rapid siltation of several reservoirs constructed since the late 1950s;
- high sediment loads in the major streams; and
- development of river terraces.

Such sedimentation may explain the current low water-holding capacity of many dams and reservoirs in Zimbabwe. For example, it has been found that out of the more than 8 000 dams with a total storage capacity of 4.9×10^9 m^3 of water, those in some of the drier parts of the country were holding only 15–25% of their capacity, even toward the end of the rainy season (Holmberg and Timberlake 1991; Moyo et al. 1991, 1993). This is due to soil erosion and sedimentation, even though there are no data to explain the exact impact. Under such circumstances, the water supply is greatly reduced and could lead to demand competition.

Despite its importance, the soil-erosion issue seems to have received limited attention in the environmental agendas of the countries in the subregion. Within the division of portfolios among the Southern African Development Community (SADC) member states, Lesotho has responsibility for soil erosion. This is in response to the country's extensive environmental problems (Moyo et al. 1993). As early as 1936, technical attempts to combat soil erosion in Lesotho were initiated in the form of contours along undulating lowlands and tree planting. These early efforts had limited success because of the forceful methods used by the colonial government.

In terms of research, there is a shortage of information on soil-erosion studies, making data-compilation difficult (Stocking 1987). There seem to have been only limited proposals on this issue at both the national and the regional levels, so numerous opportunities exist. Based on the current situation, future research efforts should focus on

- determining the relationship between drought and soil erosion;
- determining the rate of sedimentation and impact on dams; and
- assessing the role of human activities in generating soil erosion.

Rapid population growth

The dynamics of population in the region and the pressure on available water resources are a primary issue to be considered in discussing the sources of stress on water resources. During the past decade, southern African countries have experienced rapid population growth. The population of many countries has almost doubled in this same period, with no additional increase in water supply, meaning that available water resources have to be shared among competing demands.

Current estimates put Zimbabwe's population growth rate at 3%, which, if sustained, would produce a population of 15 million by 2006 and 20 million by 2016 (Moyo et al. 1993).

The effects of population on water resources are manifested in the water-demand side of the issue, for there is a serious demand to satisfy needs, as well as sectoral demands for agriculture and industry. In Zimbabwe, current water-consumption levels for the various sectors show that the bulk of water is consumed in urban areas (representing 23% of the population) and in commercial farming and mining areas (together representing 20% of the population). Communal areas (representing 57% of the population) account for a smaller portion of water consumption, as drinking water for people and livestock is supplied mostly through groundwater at point sources (Holmberg et al. 1991; Moyo et al. 1991, 1993). These figures do not indicate whether the demand of all water-user sectors is met adequately, nor do they show the level of the demand.

There is no indication that governments are attempting to evolve any policy to control population growth in Zambia or Zimbabwe, or indeed anywhere in southern Africa. No country has a clear government policy on population.

To address issues bordering on population demand for water, the DWA in Zambia, under the CMMU-WASHE program, is carrying out field studies to determine per capita water demand (Carty 1994). This is being done to determine existing water demand in relation to supply; however, more actions are needed.

A research agenda could be based on the United Nations (UN) global position on the population and resource relationship and address issues such as

- the need for countries to know their national population-carrying capacity (the capacity of the resource base to support and provide for the needs of human population);
- the need to pay special attention to critical resources such as water; and
- the need to deal with migrations that result from and lead to stress on water resources, that is, to environmental degradation.

Industrial pollution

The SADC region is not highly industrialized in comparison with Western industrial nations. However, it has great potential for industrial development, especially after passing through the long transition period from colonial domination to self-governance and economic determination. Owing to the low level of industrial activity, the impact of this sector on the environment in general and on water resources in particular has been localized.

The industrial pollution of the Kafue River, an important freshwater source in Zambia, is a good case in point (Macfoy 1994). The river passes through Zambia's copperbelt industrial complex, an area where major mining, agricultural, and hydropower schemes are located. Some of the major industries in this complex include Zambia Consolidated Copper Mines, Rokana Corporation Limited, Kafue Textiles of Zambia, Nitrogen Chemicals of Zambia, Bata Tannery, and National Breweries. Industrialization has brought with it extensive drainage and pollution problems. Because of the lack of water-disposal systems, the practice of the industries has been to discharge industrial waste directly into the Kafue River, making it perhaps the most polluted river in the subregion. The consequences of pollution on the Kafue include

- high concentration of chemicals in the water;
- reduction in the quantity of freshwater consumed by humans and animals as the quality of drinking water is increasingly affected;
- increased turbidity;
- proliferation of aquatic weeds (eutrophication);
- disappearance of aquatic animals; and
- silting of the nearby dams.

Consequently, as less water has been available, people have been forced to explore alternative sources of water, both for human consumption and for livestock.

The Zambian government has taken some action to deal with the pollution of the Kafue River. In 1985 the government adopted the National Conservation Strategy, and in 1990 Parliament passed the *Environmental Protection and Pollution Control Act* (EPPCA), providing the legal framework for the smooth implementation of the strategy. Under EPPCA, it is a serious offense to discharge effluent with levels of pollutants higher than the minimum acceptable level or to abstract water without a licence. The government also established the Environmental Council of Zambia to enforce EPPCA. In 1993, the *Pollution Control Regulation* was reinforced by the *Waste Management Regulation*, which mandates the council to restrict the handling, storage, transport, and treatment of waste. A task force was also set up to manually remove aquatic weeds in the river, but this proved unsuccessful.

Despite measures such as these taken by the authorities to control pollution, it is evident that not much has been achieved: there are still flagrant violations of the *Pollution Control Regulation*, and enforcement of the legislation has been ineffective. There is, therefore, a need to reexamine the enforcement strategy to make the legislation more effective.

Research initiatives on pollution need to investigate the effects of pollution on humans and animals, as well as the effect of soil acidity on plants. In addition, a systematic program for monitoring water quality, in line with the Hydrological Cycle Observing System for SADC (HYCOS–SADC), is needed to improve the collection and management of hydrological data.

Commercial agriculture

Commercial agriculture involves cash-crop and livestock production for trade and income generation. It may involve an intensive system, with high application of inputs to increase yield per hectare, or an extensive system, entailing cultivation of large expanses of land. There are also two types of commercial agriculture: large-scale commercial farms and small-scale commercial farms. Both use large quantities of water for irrigation and livestock to increase production.

Commercial agriculture is widely practiced throughout the region. In Zimbabwe, the total irrigated land under commercial agriculture is 14.1×10^6 ha, with 12.7×10^6 and 1.4×10^6 ha for large- and small-scale commercial farms, respectively (Whitlow 1988), about 36.2% of the total arable land. Most of the water used in the country comes from surface water impounded by dams, and almost 90% of the estimated 4.9×10^9 m^3 of water stored is used for agriculture (Holmberg et al. 1991; Moyo et al. 1991, 1993), leaving only 10% for other uses.

According to Moyo et al. (1993), the problems and constraints of agricultural production stem from the rapidly growing population and the finite boundaries of the communal areas. Demand for household arable plots leads to encroachment into grazing areas, but cattle numbers continue to rise. Overstocking is most frequently cited as an indicator of imbalance between resources and people. In Botswana the cattle population is 2.5 million, and cattle continue to be dominant in Swaziland's livestock sector.

Commercial agriculture usually involves excessive use of water, which puts strain on the available water supply. Moreover, chemical fertilizers, herbicides, and insecticides intensively applied to farmlands can be washed into watercourses by surface runoff and cause contamination. Concentrations of nutrients have apparently created eutrophication and blooming of aquatic weeds, such as *Salvinia molestica* in the Okavango River and *Eichorrnia crassipes* (water hyacinth) in the Kafue River, Harare's Lake Chibero Reservoir, and the Upper Umgusa Dam near Bulawayo (Moyo et al. 1993). Water hyacinth is already a problem in Lake Victoria and upstream in the Kagera River.

Agriculture remains a major economic activity and source of income to most countries in southern Africa. Consequently, solutions to address the environmental problems associated with commercial agriculture appear to have received little attention. In Zimbabwe, for example, the *Natural Resources Act 1941*, which was amended in 1975 and 1981, provides for the establishment of Intensive Conservation Areas, in which commercial farmers are required to undertake soil-conservation measures. However, the issues of water consumption, open-channel irrigation with its associated problems (such as mosquito infestation), and contamination of water courses from farm drainage have not been addressed. Furthermore, the problem of overstocking seems to have been left to ecological balance or the ability of the ecosystem to support the livestock. Therefore, the *Natural Resources Act* is of limited value as an instrument of natural resource management, particularly in areas where degradation is most acute and where there is no other law or policy to ensure enforcement.

The foregoing analysis suggests certain research questions. For example, the relationship between eutrophication and farm-water drainage needs to be determined. Furthermore, it is important to compare water loss by evapotranspiration through open-channel irrigation with water loss through other irrigation methods, such as the drip system. Finally, it is important to explore groundwater-recharge potentials under irrigation conditions.

Hydropower development

One aspect of water use and management in the region involves the production of hydropower. Apart from generating electricity, water impounded in dams is used for irrigation, flood control, drinking-water supplies, stock watering, and fish farming. Such integrated water management is currently perceived as key to sustainable development in several sectors of the economy, and there are possibilities for such development to continue into the future. A coordinated program for hydropower development is being implemented jointly between Zambia and Zimbabwe under the framework of the Zambezi River Authority. Some projects have been completed under the program. Several other hydropower schemes have been proposed to supply electricity to essentially all the countries sharing the river system.

In Mozambique, five schemes are in operation: Cobra Bassa (2 075 MW), Chicamba (38 MW), Mavuzi (5 MW), Lichinga, and Cuamba. At present, approximately one fifth of the potential hydroelectric power has been developed in the Cobra Bassa scheme, stage 1 of which has been completed (Moyo et al. 1993). Similarly, in Lesotho, the Lesotho Highlands Water Project is developing an estimated 1 300 GW (Moyo et al. 1993). It is evident, therefore, that the combined demand for hydropower in the region is enormous.

There seems to have been no consideration of the impact of dams on the water systems, although dams and river diversions have been known to affect both water quality and water quantity, leading to local and temporary water shortages (Keating 1993). For example, the major rivers of Zimbabwe, such as the Save and Limpopo, become rivers of sand with muddy pools joined by a trickle of brown water by the end of the dry season (Moyo et al. 1993). Lake Kyle is also reported to be suffering from dangerously high evaporation rates. These problems are further compounded by the rapid siltation of reservoirs. The impact of dams is certainly felt from the supply angle, as much strain is put on available water supplies. Resettlement schemes also cause the migration of people, who might be forced to move into marginal areas where further pressure is put on the available water supply. It is in recognition of the adverse effects of large dams on the environment in general and on water supply in particular that some pressure groups, such as the International Commission on Large Dams, were formed to monitor the construction of dams worldwide.

Actions to address the problems caused by dams have met with limited success. The National Master Plan for Water and Sanitation in Zimbabwe, for example, commissioned a special study for which a sedimentation survey of 30 dams was undertaken (Moyo et al. 1993). Furthermore, the Department of Natural Resources put forward a standing requirement that a report outlining the state of the catchment area would have to be produced by the Natural Resources Board before construction of major dams. Unfortunately, none of these bodies have the appropriate methodology, human resources, or equipment to execute such studies. Consequently, most dams are still being constructed without adequate information.

Some pertinent issues related to dams need further research. These include siltation surveys, assessments of impacts on water quantity and quality, and assessments of socioeconomic impacts on the less privileged, especially women, the elderly, and children.

Mining activities

Southern Africa is rich in mineral resources, including gold, copper, coal, iron, and bauxite, and their exploitation plays a dominant role in the economic development of the subregion. In some areas, the history of mining dates back more than 50 years (Moyo et al. 1993). Two methods of mining are commonly employed: opencast and strip.

In Zimbabwe, more than 1 000 small mines have been worked within the last century. The rate of exploitation is high, and some experts have predicted that known gold and copper ore deposits, estimated at 0.6×10^6 t, would be exhausted, at present rates, within 10 years. About 3×10^6 t of coal per year is mined at Hurange, and the estimated reserves are enormous, ranging from 4×10^9 to more than 20×10^9 t. Similarly, in Zambia, decades of intensive copper exploitation in the copperbelt have created a barren and unproductive landscape. Mining is carried out both by small mining cooperatives and by transnational companies like Rio Tinto Zinc, Lonrho, Anglo-American Corporation, and Union Carbide.

Mining activities have adverse consequences on water supply and the environment because the mining process involves the use of large quantities of water in jets to remove the overburden and wash the mineral ore. In the copperbelt, for example, the mined ore is crushed and milled with water to a very fine particle size. Chemicals are added to the pulp, which is then agitated with air in tanks (Moyo et al. 1993). Wastewater is discharged into nearby watercourses, thereby posing a serious danger of water pollution. Thus, the mining industry has the potential to cause destruction of the watershed, subsidence of land, and disruption of the hydrological balance. Studies undertaken in Zambia have shown that polluted water was responsible for the deaths of livestock in the copperbelt and that the copper content in the affected rivers was on average about 80 times higher than the acceptable level.

Although the mining industry certainly affects the supply and demand for water and puts strain on the available water resources, action to reduce the impacts of mining have been limited. Several acts on the statute book in Zimbabwe attempt to control the environmental damage caused by mining. These include the *Hazardous Substances and Particles Act* and the *Atmospheric Pollution Prevention Act*. The *Mines and Minerals Act*, however, overrides most other acts, and very few restrictions are attached to the exploitation of mining rights after a mining permit has been obtained. This makes enforcement difficult and leaves the way open for negative impacts. Consequently, there has been little or no action by mining companies in the region to rehabilitate degraded landscapes.

Future action, particularly from the point of view of research, should focus on quality monitoring to establish the possible human health effects of using contaminated water. There is also a need to investigate the impacts of mining on underground water.

Water supply and demand

Against the background of declining supply and rising demand, water availability in southern Africa is a critical issue. In examining the supply and demand situation, I shall highlight the total amount of water required by the agriculture, domestic, industry and mining, and tourism sectors.

Per capita availability of freshwater

Zimbabwe's *Financial Gazette* of 16 December 1993 contained the following item on the current global water-supply situation:

> While the amount of water in our rivers and streams catered for the population a decade ago, the time is rapidly approaching when we simply have too little of it for everyone to utilize. Water is money. Until people think of water in the same terms as money, we are headed for disaster. In the end, unless we change our way of thinking, water will be a far more emotive issue than land.

This statement aptly describes the prevailing situation in southern Africa. Available data already indicate that the annual per capita availability of freshwater declined significantly from about 2×10^4 m^3 in 1950 to about 1×10^4 m^3 in 1980 (IDRC 1994). In theory, sufficient water is available to meet demand, but it is unevenly distributed in time and space (Moyo et al. 1993). In Zimbabwe, the current water-consumption levels for the various sectors show that most water is consumed in urban areas and on commercial farms; the communal areas account for a smaller proportion.

Some attempts have been made to quantify the water demand. The CMMU of the DWA in Zambia has carried out field studies to determine per capita water demand (Carty 1994). Based on observations made by various project personnel, CMMU has established a per capita requirement of 20 L/d in rural areas. This is considered "adequate" in the context of current practices, but this is not to say that the amount will not increase in future. Rather, 20 L/d is considered the minimum adequate amount when the actual availability of rural supply is taken into account, given current practices and traditions. The requirement in urban areas is double that of rural areas. Using the UN-established per capita requirement of 40 L/d for urban residents, I calculated a 30 L/d average requirement. Based on this figure, it should be possible to estimate the daily water requirement in the population. It should also be possible to project future water requirements in any given year.

According to Grizic (1980), the annual surface-water flow available for use in Zimbabwe is 80×10^9 m^3. If this quantity is fully utilized, it would suffice for a population two to three times greater than that of 1990, even allowing for a doubling in per capita consumption. Thus, total hypothetical availability should not become a serious constraint for at least 30 years, by which time improved pumping technology, repurification of urban water, and more economic irrigation methods should have become more widely used.

This estimate essentially suggests that the available water in absolute terms has the capacity to satisfy requirements at the current level and even the potential to meet future rise in demand. Nevertheless, issues such as management problems, pollution, and degradation of water resources do not appear to have been considered. Certainly, these factors seem to account for the inability of water authorities to meet current demand.

In some countries of the subregion, the responsibility for providing water is shared by several institutions and government departments. In Zambia, seven ministries, departments, and local authorities share responsibility for water. Some of them are concerned with policy formulation, and others deal with actual delivery of services to consumers (Chiwala 1994).

A similar institutional arrangement exists in Zimbabwe, where two authorities are responsible for the provision of water: the Department of Water Development (DWD); and the District Development Fund (DDF) in the Ministry of Local Government, Rural and Urban Development. DWD is in charge of hydrogeological surveys, including construction of large dams, whereas DDF is in charge of the construction of small- and medium-scale dams and deep wells. Shallow wells are under jurisdiction of the Ministry of Health and Child Welfare. Both DWD and DDF undertake the siting and sinking of boreholes in designated areas; however, maintenance is done by DDF.

Agriculture

Sectorally, agriculture is the biggest water consumer. Commercial agriculture is largely responsible because it uses large-scale irrigation systems. Because southern Africa lies in a region that receives inadequate rainfall to support large-scale rainfed crop production, irrigation provides the only alternative method of sustaining the agricultural system. As noted, almost 90% of the total water impounded in dams is used for agricultural purposes. This represents a gross overallocation of a critical resource and calls for a review of the water-allocation strategy employed in the region to ensure that the other sectors get a fair share of available water supplies.

Domestic

Water demand for domestic uses has grown steadily in proportion to population growth. Access to clean water for domestic purposes, such as drinking, cooking, washing, personal and environmental cleansing, and sanitation, is constrained by management problems. Consequently, demand does not match the supply.

Industry and mining

The industrial and mining sectors together share about 5% of the available water supply. Although the region's industrial sector is not yet well developed, the mining industry is a major force in the economic development of the countries.

The types of industries, together with the production process, determine the extent to which water is used. Generally, most of the industries are agriculturally based and include food processing, canneries, textiles, and breweries. More than 80% of the water used in these industries is used for cleansing, rather than for the actual manufacturing process. In the case of mining, water is used in opencast mining and in washing the mineral ore before it is processed.

Tourism

Tourism, a growing industry in the region, also involves on-site use of water. Kariba Dam (the largest engineered water reservoir in Africa), Livingstone Falls, Lake Victoria, and the Okavango River are important tourist centres. The extent of tourist activity in these areas depends especially on the preservation of the natural environment.

Management actions

Many water-resource programs and projects have been proposed in the region to cope with water stress and water demand and at the same time to protect and improve the quality of the environment. Some of these have already been completed or are ongoing.

Community level

Two case studies, one from Zambia and the other from Zimbabwe, illustrate community action in water-resource management.

The Kanyama water project (Zambia)

The Kanyama community is a squatter settlement within Lusaka. Before the Kanyama water project was initiated in 1988, residents depended on water wells and they felt that the city council should provide them with piped water. Subsequently, the council was approached with the proposal, but it had no funds. At this juncture, Human Settlements of Zambia (HUZA), an NGO, took up the challenge because Kanyama was an area that needed upgrading (Chitondo 1991). WaterAid-UK provided funds to HUZA, and the project started in 1987 and was completed in 1989. Project planning was done jointly by the community leaders of Kanyama, HUZA personnel, and the staff of Lusaka Urban District Council. Implementation was undertaken entirely by the residents themselves, with HUZA playing only a coordinating role. On a self-help basis, the community also moulded concrete blocks for building a wall around the borehole site, with staff from the NGO providing technical advice. HUZA provided the community with instructions for guarding and maintaining the project. Other actors in this project were the Zambia Electricity Corporation, which installed the transformer, and the Lusaka Water and Sewage Company, which took over the maintenance. It was expected that the project would provide a permanent solution to the problem of water supply in the community. Indeed, this expectation has been realized, and the community now has a regular water supply.

The Chisvoteso water project (Zimbabwe)

The Chisvoteso water project dates to 1984, when the community of Seke decided to embark on a water project as part of a large vegetable-growing project. The objective was to provide enough water for domestic consumption and for watering gardens and fruit trees that were intended to generate income (Mashongamhende 1991).

The project was initiated by the Association of Women Clubs (AWC). In 1987, further assistance came from Africare and Environment Liaison Centre International, through AWC, which enabled the women to sink a well and install a handpump. Implementation of the project was accomplished primarily by the women themselves, with assistance from AWC staff in technical areas.

The project was successfully completed in 1987. The well has made it possible for project members to grow vegetables intensively the year round and raise their family incomes and living standards. Now other provinces in Zimbabwe are benefiting from this experience and embarking on similar projects.

National level

Most governments in the region have made little progress in developing concrete programs to address critical problems in water-resource management. Water-resource management is undertaken by various ministries and departments, often in an uncoordinated manner, leading to duplication of effort and total neglect of some aspects of the environment. Nevertheless, some national programs in this sector are being initiated, as is illustrated by Zambia and Malawi.

Government effort to improve sector performance (Zambia)

Recognizing the need to achieve long-term improvements in the performance of the water-supply and sanitation sector and improve its attractiveness for foreign investments, the Zambian government set up an interministerial committee, the Programme Coordination Unit (PCU), in 1993. The PCU comprises all institutions involved in the provision of water-supply and sanitation services (Chiwala 1994).

PCU is a policy-making body that targets the entire water sector. Its main objective is to advise government on the reorganization of the water-supply and sanitation sector. Its tasks include

- recommending policy reforms for the water-supply and sanitation sector;
- defining the responsibilities of ministries and organizations in the water-supply and sanitation sector;
- determining and recommending necessary reforms and reorganization of the sector;
- proposing the creation of a framework for planning, development, operation, and maintenance of the infrastructure in the water-supply and sanitation sector, to encourage and optimize donor support; and
- proposing reforms to strengthen various institutions with responsibilities in the water-supply and sanitation sector.

PCU, through its executive arm, the Water Sector Development Group, produced a policy document on water-sector reorganization, which went before Parliament for approval. Future proposals for the water sector are

- transformation of PCU into the National Water and Sanitation Council; and
- creation of council-owned regional companies to assist councils in the delivery of water to peri-urban and rural areas, in collaboration with the Ministry of Local Government and Housing.

The Zambian initiative represents a step in the right direction. It has the potential to improve the water supply and address other issues in the water sector. However, the approach lacks public participation, and PCU failed to link grass-roots concerns to technical expertise in its development of a national plan.

Institutional arrangement for water management (Malawi)

Development action programs in Malawi are arranged in line with the Malawi Congress Party's political structure. Development area committees oversee and supervise the planning and implementation of various development projects, including water. They are organized according to the political administrative division of the country, as outlined by Moyo et al. (1993):

- The National Development Committee, chaired by the President or a representative, reviews national development programs.
- The three regional development committees (one in each region), chaired by regional administrators, prepare and review development programs that have a regional impact.

- The 24 district development committees (one in each district), each chaired by a district commissioner and with representatives drawn from economic sectors in that district, discuss development issues affecting their district.
- The 192 area action groups represent traditional authorities, who chair the meetings and supervise any projects arising from them.
- Village action groups are found in each village. Village headmen chair the meetings and supervise any development activity arising from them.

The Malawi structure is a policy arrangement designed to enhance water supply and manage demand from the village level. The broad-based approach is desirable for grass-roots contribution to water development and has the potential to improve water supply if bureaucratic bottlenecks are removed. There is also a need to create water subcommittees at the various levels to give closer attention to critical water-resource issues in the country.

Regional level

There are several water-resource projects in the southern African subregion. These projects include

- Zambezi River System Action Plan (ZACPLAN);
- Regional Hydrological Assessment Project;
- HYCOS–SADC;
- Regional Hydroelectric Hydrological Program (SADC Project);
- Water Resources Planning for SADC (SADC Project);
- SARP region water-sector assessment;
- Southern Africa FRIEND Project; and
- Rationale for Water Sector Capacity Assistance.

Two of these projects are reviewed in the following sections.

Zambezi River System Action Plan

In 1992, the treaty establishing SADC came into force. Before this, a major political challenge facing the countries was managing the international watercourses in an integrated manner to avoid conflicting interests. Integrated management comprises all the tasks required to provide water of acceptable quantity and quality to satisfy the needs of all users in an environmentally and economically sustainable manner. This approach is deemed to ensure optimal use of water resources for economic and social development. Against this background the various regional water-resource action programs were mapped out under the framework of SADC.

The overall target of ZACPLAN is to ensure that the shared resources of the Zambezi River are used in a manner that guarantees maximum long-term advantages to all participating states. The plan, adopted by SADC in 1987, consists of 19 projects (ZACPROS). Category 1 projects are selected for immediate implementation as a matter of priority. Eight such projects were elaborated and presented for funding in 1989. The specific objectives and targets of ZACPROS are to

- create an inventory of existing and potential development projects, evaluate the environmental impact of major projects, and initiate a basin-wide information exchange;
- develop regional legislation necessary for the management of the Zambezi and minimum national legislation required by riparian states for enforcement;
- develop a basin-wide unified monitoring system related to water quality and quantity;
- develop an integrated water-resource-management plan for the Zambezi River basin, create a relevant water-quality and water-quantity database, and review all sectors that benefit from or affect water-resource development projects in the basin; and
- develop and adopt a management simulation model, simulate the various development scenarios in the basin, and present an integrated water-resource-management plan.

The SADC Environment and Land Management Sector (ELMS) was entrusted with executing and coordinating the program, assisted by a committee representing all member states. This later became the ELMS Water Resources Subcommittee, responsible for advising ELMS on regional water resources issues.

The ZACPLAN constitutes a pragmatic action plan that could address critical transboundary water-resource issues for the Zambezi River basin. It must be pointed out, however, that to succeed the scheme must seek and join the concerns of riparian communities in the plan and safeguard against political interest.

It is too early to evaluate the success of ZACPROS because the time from adoption to funding and implementation has been short. Moreover, more than half of the projects have not been elaborated or funded. Nevertheless, with the spate of reorganization and coordinated action taking place between the countries in the region, there is indication that the mandate of ZACPROS could be fulfilled.

SARP region water sector assessment

The SARP region water sector assessment was initiated by SADC and funded by USAID. It covers the SADC countries, including South Africa and Zaire. Like ZACPROS, the SARP project was adopted in 1987. However, no definite time frame was set for its implementation.

The objective of the scheme is to determine the organizational structures that exist within the southern African region and to consider how water resources can be effectively coordinated on a regional, as opposed to national, basis. Phase I will gather information on all existing major water-resource schemes in the countries and identify all ministries, agencies, authorities, and donors involved, with emphasis on regional institutions. Phase II will evaluate the information and recommend interventions in infrastructure and institutional development that would foster regional water-resource sharing and development. It will also identify possible infrastructure development projects that might be considered by the regional water entities.

Current institutional reorganization in the water sector in many countries in the region conforms with the intended actions of this project. A similar institutional setup is also desirable at the community level. A well-coordinated institutional framework could enhance the further development of water-supply infrastructure in the region.

Conclusion

Based on the information gathered and the analysis in this paper, it is certain that the major freshwater sources in the region are under stress from a number of factors. These include drought, rising population, pollution from industries, and mismanagement leading to wastage. The current and projected water demand is on the increase, and there are calls for action from all concerned. Several actions are being proposed, others are ongoing, and some have been completed at community, national, and regional levels to cope with the stress and meet rising demand.

However, improving the quality and quantity of the water supply to meet rising demand is still a major problem facing the region and is likely to remain so in the future. Holmberg et al. (1991) and Moyo et al. (1991, 1993) proposed three basic options:

- reallocate available water to meet new emerging demands;

- increase supplies by tapping additional groundwater, building dams, or transferring water from surplus to deficit areas; or
- stretch supplies through conservation, recycling, more efficient use and pricing policies.

The first option is the subject of review under the new water-sector master plans being developed by the various governments and, at the regional level, under the water-resources agenda of SADC.

The second option provides an obvious alternative. It would involve an inventory of water resources, not previously fully undertaken in the countries, particularly for groundwater. Adequate management of water resources might only be possible if data are obtained on available water stocks. In the case of building dams, a number of constraints, such as the adverse effects of dams on the environment and the low status of most user populations should be considered. Interbasin water transfer is often a cost-effective venture and is currently being done in western Zimbabwe.

The third option, although widely advocated, would present problems, notably to large-scale commercial irrigators. This approach would, however, be beneficial to communal farmers, as they are known to face problems of soil degradation and erosion, which cause loss of topsoil and blockage of streams through siltation.

Based on the analysis, additional options are as follows:

- The development of efficient rainwater-harvesting systems and storage of runoff floodwater could easily be adopted at the family and community levels. These projects would involve construction of medium- to large-scale water reservoirs, including earth dams to store floodwater within a catchment area during rainstorms. This option is desirable because an estimated 20×10^9 m^3 out of 227×10^9 m^3 of storm runoff generated in the region reaches the rivers (Holmberg et al. 1991). If harnessed, this runoff could provide short-term relief to several communities.
- Desalinization of seawater, although expensive, might be a last resort. It would provide a lasting solution, because use would be made of the oceans surrounding the region.

Research opportunities

Proposed research programs in the water sector are based on the identified sources of stress. The main research questions that should be addressed are presented here:

1. Undertake a comprehensive field survey of boreholes and determine the stock of surface water and groundwater to develop a basis for future planning and allocation of water for different uses.

2. Document drought incidents, trends, and impact and identify long-term strategies, including early-warning systems, to cope with the phenomenon of drought.

3. Determine the relationship between drought and soil erosion, the rate of sedimentation and siltation and the impact on dams, and the role of human activities in generating soil erosion.

4. Evaluate the population-carrying capacities of the water systems and the effects of migration, paying special attention to water as a critical resource.

5. Assess the impact of pollution on humans and livestock and of soil acidity on plants. A systematic program to monitor water quality and quantity is required to improve the collection and management of hydrological data.

6. Assess the relationship between farm-water drainage and eutrophication.

7. Compare the extent of water loss through open-channel irrigation with that through other methods, such as drip and sprinkler irrigation. Furthermore, the groundwater-recharge potentials in areas under irrigation should be explored.

8. Carry out siltation surveys, assessments of the impacts of dams on water quantity and quality, and assessments of the socioeconomic impacts on the vulnerable groups, especially women, children, and the elderly.

9. Monitor the impact of mine-water drainage on public health and the environment. Investigate the impacts of mining on groundwater reserves after the application of strip-mining methods.

References

Arntzen, J.W.; Veenendal, E.M. 1986. A profile of environment and development in Botswana. NIR Free University, Amsterdam, The Netherlands.

Carty, D. 1994. Defining coverage parameters for rural water supply and sanitation in Zambia. CMMU, DWA, Lusaka, Zambia. Discussion Paper No. 4.

Chakela, G.K. 1981. Lesotho: soil erosion and reservoir sedimentation. Scandinavian Institute of African Studies, Uppsala, Sweden. Report No. 54.

Chitondo, E.M. 1991. Human settlements of Zambia and integrated water projects. Community participation and water supply. AWN, Nairobi, Kenya. pp. 83–85.

Chiwala, B. 1994. Reorganization and institutional development in the water and sanitation sector. Paper presented to DWA/IRC July 1994 Workshop on Community Participation, Livingstone, Zambia.

Connolly, D. 1991. Rural water supply in Zimbabwe: overview. *In* Community participation and water supply. AWN, Nairobi, Kenya. pp. 86–96.

Holmberg et al. 1991. Defending the future: a guide to sustainable development. Earthscan Publications, London, UK.

Keating, M. 1993. Combating desertification and drought. Agenda for change. UN Centre for Our Common Future, Geneva, Switzerland. pp. 21–22.

Macfoy, C.Z. 1994. Pollution of the Kafue River. Paper presented at the Regional May–June 1994 Workshop on African Freshwater Resources: Challenges for the 90s and Beyond, Nairobi, Kenya.

Mashongamhende, R.Y. 1991. The Association of Women Clubs and water supply in Zimbabwe. *In* Community participation and water supply. AWN, Nairobi, Kenya. pp. 92–95.

Moyo et al. 1991. Zimbabwe's environmental dilemma, balancing resource inequalities. Harare, Zimbabwe.

——— 1993. The southern African environment: profiles of SADC countries. Earthscan Publications, London, UK.

Stocking, M. 1987. A methodology for erosion hazard mapping in the SADC region. SADC soil and water conservation program. Maseru, Lesotho, No. 9.

Whitlow, J.R. 1988. Land degradation in Zimbabwe: a geographical study. Department of Geography, University of Zimbabwe, Harare, Zimbabwe.

Improving Water Supply Systems in Rural West and Central Africa

Bassirou Diagana
Hydrogeologist, Ouagadougou, Burkina Faso

Introduction

The International Drinking Water and Sanitation Decade (1980–90) had the objectives of providing water and sanitation to rural and urban communities in Africa and involving communities in the management of their water resources. However, some of these objectives have not been met. Barely 45% of the per capita requirements for water are met by the 10–20 L/d available. New approaches to improving access to water have included the following:

- equipping villages and urban centres with water points;
- improving the supply of equipment for existing water projects; and
- creating user awareness of efficient water-resource management.

Sources of water supply

Surface water

Lakes, rivers, and streams are the main sources of surface water. Because of its accessibility, surface water is more widely exploited than groundwater, especially in the wettest parts of Africa. Except in periods of severe drought, the big rivers of West and Central Africa — the Senegal, the Niger, and their great tributaries, the Chari, the Logone, and the Bandama — discharge millions of cubic metres of water each year. However, this water is not fully exploited by riparian countries.

Groundwater

Groundwater is subsurface water. Aquifers are permeable formations, either consolidated (for example, sandstone) or unconsolidated (for example, sand), that store and transmit groundwater in sufficient quantities to supply wells. Aquifer layers can be continuous, discontinuous, or mixed. Table 1 lists the formations associated with each type of aquifer layer.

Continuous aquifer layers are the most important in many respects: potential, extension, renewal, and discharge rate. These aquifer layers, found in unconsolidated formations, are largely sought to supply water to big urban centres, agriculture, and industry.

The discontinuous aquifer layers are localized and linked to fracturation systems in both arid and semi-arid zones, as well as in humid tropical zones. These aquifer layers are found in crystalline and crystallophyllian formations.

Table 2 shows that Cape Verde and Sao Tome and Principe, which have a small land area, have only one type of aquifer layer. Most of the other countries have the three types. The area of each aquifer type as a proportion of total land area suggests the methodology to be applied for digging water holes and the typology of works to be adopted.

Table 1. Main hydrogeological units.

Type of aquifer layer	Formation
Continuous	Sand
	Clayey sand
	Sandy–clayey clay
	Clay and marl
	Limy clay
Mixed	Clayey sandstone
	Sandstone
	Limy sandstone
	Limestone
	Marly sandstone
	Dolomite
Discontinuous or local	Schist
	Quartzite
	Conglomerate
	Volcanic rock
	Dolerite
	Crystallophyllian rock
	Granite

From 1976 to 1982, the Inter-African Committee on Water Studies (Comité interafricain d'études hydrauliques) published a series of hydrogeological maps at scales from 1 : 1 000 000 to 1 : 1 500 000. These maps have greatly contributed to knowledge of water resources, especially those in crystalline bedrock. The maps have made it possible to know with precision geological contours, exploitable reserves, conditions of exploitability, and groundwater quality.

Table 2. Area of each aquifer type as a proportion of total land area.

State	Land area	Continuous		Mixed		Discontinuous	
	(km^2)	(km^2)	(%)	(km^2)	(%)	(km^2)	(%)
Benin	112 622	11 262	10	13 515	12	87 845	78
Burkina Faso	274 200	10 968	4	32 904	12	230 328	84
Cameroon	475 442	95 088	20	85 580	18	294 774	62
Cape Verde	4 033	—	—	—	—	4 033	100
Central African Republic	622 984	112 137	18	49 839	8	461 008	74
Congo	342 000	92 340	27	218 880	64	30 780	9
Côte d'Ivoire	322 463	12 899	4	—	—	309 564	96
Gabon	267 667	2 677	1	139 187	52	125 803	47
Gambia	11 295	9 375	83	1 920	17	—	—
Ghana	238 537	14 312	6	95 415	40	128 810	54
Guinea	245 857	2 459	1	66 381	27	177 017	72
Guinea-Bissau	36 125	14 450	40	21 675	60	—	—
Equatorial Guinea	26 017	1 041	4	1 041	4	23 935	92
Liberia	111 500	2 230	2	—	—	109 270	98
Mali	1 240 710	645 169	52	459 063	37	136 478	11
Mauritania	1 030 700	515 350	50	237 061	23	278 289	27
Niger	1 267 000	836 220	66	329 420	26	101 360	8
Nigeria	923 768	230 942	25	138 565	15	554 261	60
Sao Tome and Principe	955	—	—	—	—	955	100
Senegal	201 400	128 896	64	56 392	28	16 112	8
Sierra Leone	71 740	8 609	12	7 174	10	55 957	78
Chad	1 284 000	770 400	60	282 480	22	231 120	18
Togo	56 000	4 480	8	8 960	16	42 560	76

Computer-assisted cartography has made it possible to create topical maps at more useful scales. At a scale of 1 : 500 000, such maps can present groundwater data meeting three essential specifications: accessibility, or works cost; security, or exploitation cost; and productivity, or exploitation duration.

Legal and institutional framework

When surface water, such as a river, is shared by more than one country, conflicts can arise when one of the partner states decides to exploit the common water source by diverting it or erecting a dam. Obviously, a consensus is needed. The Manatali and Diamana dams provide examples of this.

Groundwater, especially the aquifer layers of the big sedimentary basins, always crosses national boundaries. In North Africa, these layers are intensely exploited for agro-industrial purposes. In sub-Saharan Africa, the volumes of water drawn from the great basins are insignificant compared with the existing reserves and replenishment volumes. That is why the exploitation of these groundwaters is less constrained than the exploitation of the big watercourses that often feed these underground reserves.

A common resource must be regulated by legal provisions to facilitate the cohabitation of beneficiaries and ensure its efficient use and its lasting exploitation for the benefit of future generations. These provisions are of two kinds: those regulating national affairs and those regulating interstate affairs. The application of national regulations is a matter of sovereignty and also takes into account national policy objectives. The application of interstate regulations usually requires talks between the partner states and may require these countries to give up a part of their sovereignty. This often makes it difficult to apply interstate regulations.

Conclusion

Studies have indicated the need for rational exploitation and management of water. Whether water is provided free of charge or distributed according to a policy of participation, the potable water supply in urban and rural areas has been inadequate. The economical character of water is no longer in question. It is up to the governing authorities to reverse the trend that gives water supply only a social value and to adopt for rural areas the water-management strategies that work in urban centres. The full responsibility for managing water points should then be entrusted to communities and local operators.

PART III

Special Issues

Water Supply and Management in Rural Ghana: Overview and Case Studies

Kwame Karikari

Senior Lecturer, School of Communication Studies, University of Ghana, Legon, Ghana

Introduction

This paper examines the extent to which government, implementors, and users are adopting different but interrelated mechanisms to deal with water stress in Ghana. It analyzes the policy frameworks being adopted to develop alternative sources of water to meet the needs of different users, the institutional frameworks used in implementing these policies, and the participation of users at the community level.

Water resources

As is the case in most parts of sub-Saharan Africa, water demand far outstrips supply in Ghana. The main sources of water for households are piped supply from treated water sources; untreated piped water from groundwater sources; shallow boreholes; wells; and ponds, springs, lakes, rivers, and streams.

Ghana has a tropical climate. There is a wide variation of rainfall, influenced by the southwest monsoon. Mean rainfall varies from 2 000 mm in the southwest coastal area to about 850 mm in the eastern coastal area and 1 000 mm in the north.

The largest river in Ghana, the Volta, with two huge hydroelectric dams, has a catchment of 165 700 km^2 within the country. Volta Lake, behind Akosombo Dam, is about 300 km long and traverses the centre of the country. The other rivers are all in the south and southwest and drain about one third of the country. However, because of heavy rainfall in these areas, the rivers make up about 50% of the internal runoff. The major rivers in the north, such as the Black Volta and the White Volta, and their tributaries are perennial. The rest are dry during the long dry season.

Sources of water supply

Urban communities get most of their water supply from rivers at dams and diversion structures. Most of the surface water has to be treated to meet health standards. Surface-water resources can probably serve all urban needs for the foreseeable future through parallel programs of development and conservation.

The rural communities rely on groundwater, provided that it is available in sufficient quantities, is reliable throughout the year, and does not require treatment. The quality of groundwater is generally good, although in some locations the water contains iron and manganese deposits. Groundwater from shallow wells near streams and springs has always accounted for a large share of the potable-water supply in rural communities. Any site with static water levels of less than 10 m is a possible source of groundwater.

Two types of wells are common, namely, hand-dug wells and drilled wells (boreholes):

- Hand-dug wells generally have a large diameter and are constructed using simple tools, such as pickaxes and shovels. Their depth ranges from 5 to 20 m. They can be lined with concrete cast on site, precast concrete rings, rock, or concrete blocks. They serve about 200 people.
- Boreholes are hand- or machine-drilled wells. Hand-drilled wells have a small diameter and are sunk using special tools, such as bits or augers. Machine-drilled wells are typically 100–200 mm in diameter and are sunk using relatively sophisticated equipment powered by diesel or electric motors. Machine drilling is suitable for depths of up to 50 m, but the depth depends on the power of the rig and the geological conditions. The number of people that can be served by a drilled well depends on the capacity of the handpump, but it is typically about 300.

Water is found in most wells, even in dry regions. The success rate of boreholes depends on the rock formation and varies from region to region. The national success rate of drilling is about 70% and ranges from as high as 95% to as low as 30%.

In 1993 a machine-drilled well with a handpump cost 4.7 million GHC (in 1996, 1 580 cedi [GHC] = 1 United States dollar [USD]), whereas a hand-dug well cost 477 000 GHC with a bucket system or 795 000 GHC with a handpump. Therefore, per dollar of investment, four to seven times more people can be served by a dug well.

For rural communities, a hand-dug well has several advantages. The use of simple hand tools for construction provides employment for local artisans, serves as a focus for community mobilization, and could be opportune for educating the villages on water-related health and social issues.

Traditional wells constructed by local artisans, however, have some limitations. In the absence of hydrogeological surveys (or when the surveys are poorly done), wells may be sited in locations with shallow water tables. In some cases, the simple hand tools used in construction may not be able to penetrate below the water table. The result may be that water is not available year-round. Poor design of the apron and the well platform (they are often left open) exposes the wells to contamination. They can also be a hazard to small children, who may accidentally fall in.

Some of these concerns could be addressed if the communities construct the wells under the supervision of technical personnel.

Water policy and program

According to Population Action International (1993), per capita available renewable freshwater in Ghana in 1955 was 9 204 m^3. By 1990, it had declined to 3 529 m^3. This is about double the amount defined as the upper limit for water stress, but it is estimated that by 2025 per capita renewable freshwater will dwindle to 1 400 m^3, well within the zone of water stress.

The Ghana Water and Sewerage Corporation

The Ghana Water and Sewerage Corporation (GWSC), a parastatal under the Ministry of Works and Housing, is the only institution with authority for developing, operating, and managing water supplies for domestic and industrial uses and for establishing, operating, and controlling sewerage systems. GWSC does not receive government budgetary allocations for recurrent expenditures and must raise funds from consumers and users to meet its overhead and maintenance costs. In urban centres, GWSC has been the principal agency for the development of water programs in recent times; in rural communities, the parastatal, donor agencies, and nongovernmental organizations (NGOs) have made significant interventions in the development of water systems.

GWSC has its national headquarters in Accra, offices in each of the nine regional capitals, and regional stores and maintenance centres in many of the 110 administrative districts. Of its 4 500 staff in 1993, fewer than 50 dealt with rural water matters, yet the corporation is responsible for maintaining the 6 600 drilled wells fitted with handpumps that serve rural communities.

The Department of Community Development under the Ministry of Local Government, which is traditionally responsible for local mobilization and education for development projects, has a technical unit experienced in the construction and maintenance of water facilities and in the construction of roads and buildings. The department has implemented water programs in the past.

District assemblies (local elected administrative structures or authorities) have also developed small water projects, although they mainly supervise GWSC water projects established by ESAs or externally funded NGOs.

National policy and programs

The government policy has been to provide potable water to consumers at the least cost. Communities with more than 2 000 people were entitled to piped water; those with between 500 and 2 000 people, to boreholes with handpumps; and those with fewer than 500 people, to hand-dug wells with buckets. However, GWSC found that extending piped water to rural communities was expensive.

By the early 1970s, there were 208 water-supply facilities under GWSC authority, mainly piped systems. By 1989, 93% of communities with more than 5 000 inhabitants had access to piped water. In rural communities, 20% of residents had piped water, and 30% relied on drilled wells with handpumps or dug wells. Although access or supply increased from the late 1960s to the early to mid-1970s, by the late 1980s an overall decline had set in.

Neither the government-subsidized operations nor consumer tariffs were based on the costs of maintenance and replacement. Instead, they were based on the community's ability to pay. A 1991 GWSC study found that, of the 208 piped water systems, 77 in the rural areas could not meet direct operating costs (GWSC 1993). Estimates showed that even an increase of 500% in prevailing tariffs would not meet the costs. Thus, in addition to general national economic decline, by the early 1980s most of the water systems under the corporation's management had run down. By the end of the International Drinking Water Supply and Sanitation Decade in 1990, only 49% of Ghanaians had access to clean water supplies.

A joint study by GWSC and the United Nations Development Programme (UNDP) in 1993 found that urban water-supply coverage had dropped from 93% to 76%. Rural water supply, however, had increased, reaching 46% of the population as a result of a number of interventions by GWSC, ESAs, and NGOs. At the end of 1993, GWSC maintained 6 574 wells fitted with handpumps in all 10 regions. Donor and development agencies altogether had established 8 695 wells fitted with handpumps in rural communities in all the regions.

In 1987, the Rural Water Division was created, and a 5-year (1987–92) program was proposed that would involve sinking 10 000 hand-dug wells in more than 7 500 rural communities with fewer than 500 people. The goal was to reach more than 2.4 million people. The program planned to drill 6 000 wells with handpumps for more than 1 000 villages of between 500 and 2 000 people, reaching a population of 1.1 million. The objective was to improve drinking-water sources in 60% of the rural areas.

User tariffs

The implementation of structural adjustment programs (SAPs), based on free-enterprise principles and the near-total withdrawal of state subsidy for social services and utilities in favour of privatization, has shaped GWSC's more recent policy of economic tariffs for users. Since 1985 user tariffs have risen by 10% for industrial and domestic use in urban and rural areas. Under the SAPs, measures taken to increase coverage and efficiency included

- revising water tariffs to ensure self-financing;
- restructuring the service to introduce regional and district autonomy in management of the services;
- assigning the ownership and maintenance of water-supply facilities to communities; and
- creating a national coordinating body for water and sanitation in rural areas.

Tariffs were first introduced in the rural areas when external agencies installed potable-water systems. The tariffs have not been well received by the communities, and many have been unwilling or unable to pay, leading to disconnection. According to GWSC, very few communities with handpumps are unwilling or unable to pay their bills. Of the 6 574 pumps GWSC maintains, 903 (14%) are out of service for mechanical reasons. Maintenance problems associated with lack of training or repair kits, as well as poor operational handling, particularly by children, have resulted in breakdowns, and many pumps need major part replacements or complete replacement.

Wherever payment default has resulted in disconnection, the impact on the communities is harsh and aggravates the water situation. With the introduction of handpumps and better quality water supply, communities have tended to neglect the maintenance of traditional water systems. In Ashanti, for example, the disconnections occurred in the dry Harmattan season, when traditional water sources are normally dry.

Rural communities with handpumps owe more than 476 million GHC, an average of 73 000 GHC per pump. The problems, which are likely to persist, involve maintenance of the systems in low-income communities. In the past, there has been no financial obligation associated with water supply, so it was taken for granted that there would be no charge for maintenance. The result is that many people will continue to depend on their traditional sources.

The communities do not see any justification for payment of tariffs. Many assume that it is the government's obligation to provide them with water. Some are willing to pay for the cost of equipment only. The argument is that, once installed, the handpumps do not consume any resources or inputs such as fuel, nor are there any renewable processes involved in purifying or improving the quality of the underground water.

New policy perspectives

In 1991, a national rural water-supply and sanitation conference was held at Kokrobite, near Accra, attended by ESAs, major donor agencies, government departments involved in rural development, and NGOs in the water sector. The conference made the following recommendations (World Bank 1991):

- Promote and foster community management of services.
- Give district assemblies a central role in supporting community management.
- Give government a role in promoting provision of service.
- Give the private sector a role (for the first time) in the provision of goods and services.
- Initiate demand-driven programs with self-selection and commitment by communities to enhance sustainability (in other words, communities would have their water resources developed if they could afford the costs of establishment and maintenance).
- Focus on women as users and on the active involvement of women as planners, operators, and managers of community systems.

An important decision was to establish within GWSC an autonomous community water-supply and sanitation department. However, it was not until 1994 that the department was finally set up with a director, staff, and office.

Today the priority directions of GWSC are

- to recover ground lost in urban water supply from the current 76–93%; and
- to upgrade the corporation's new department of rural water, first, to a semi-autonomous body and, later, to a fully autonomous body so that it can concentrate on accelerating rural water development.

The accelerated program in the new priorities will be based on demand by recipient communities and on their capacity to afford and manage the systems, especially their ability to pay operation, maintenance, and replacement costs. Improvements to the water supply will be based on commercial tariffs because of the prevailing economic policy minimizing state responsibility in social services. In rural areas, GWSC's aim is to convert water systems to community-managed services over the next 5 years. The corporation plans to install water-storage tanks at vantage points within all 110 administrative districts of the country as a principal element in its program toward creating what it calls "water security."

Parliament's recommendations

In November 1993, Ghana's Parliament appointed a Committee on Works and Housing to study in detail the problem of safe drinking water, especially in the rural areas, and make recommendations. The committee's report acknowledged that "there are communities that either cannot, or will not pay the maintenance costs, and will resort to the use of muddy water or polluted streams that may endanger their health" (Parliamentary Committee on Works and Housing 1994).

As a result of the committee's report, Parliament made the following recommendations:

- Parliament formally entrusts the district assemblies with some of the responsibility for water supply and management in their communities. In the dry season (December–April), several communities, particularly in the savanna zones, experience severe perennial water shortages because of the drop in water levels. The legislature has strongly suggested that the district assemblies and GWSC provide standby water tankers, which will sell treated drinking water at affordable prices. For cooking, washing, and other domestic uses, the communities can use their traditional streams or dams.

- Parliament proposes damming streams that do not dry up and are close to communities for large arid areas where water is scarce. The dammed water may be treated, or the people may be told to boil or filter it before drinking.

- Parliament proposes "huge underground tanks" in some areas to store rainwater in very large quantities for communities.

- Parliament asks district and metropolitan assemblies to set up "water task forces" to check wastage and pollution of valuable water by children, careless adults, and animals.

- Parliament asks GWSC to design simple and cost-effective underground tanks that could be used in homes, school buildings, clinics, and other public buildings to collect rainwater for drinking.

Program for community ownership and management

Lack of community participation has led to poor operation, maintenance, and sustainability of the water projects. This is mainly because of inappropriate technology, incorrect location of supply systems, lack of affordability, and lack of social acceptability because of "poor" or "wrong" taste of the new water supply or the presence of minerals. In some cases an inadequate survey led to siting systems where mineral content has been detrimental to tooth development in children, as with a community in the Upper West Region. However, it is evident that communities could control and manage their systems and make them work efficiently. The proposition is for communities to take greater responsibility in the financial outlay for the development of the projects and recover much of the cost of establishment and maintenance of the supply systems.

According to the United Nations International Children's Emergency Fund (UNICEF), a community should provide between 5 and 10% of the capital cost of facilities. GWSC proposes that communities provide labour for the construction of hand-dug wells, and the corporation would provide technical assistance and training for maintenance. In a scheme to provide 1 600 hand-dug wells over the period 1991–95 and safe water for 400 000 people in settlements with fewer than 500 people, UNICEF proposes to support these communities with material supplied through GWSC (Government of Ghana and UNICEF n.d.). For instance, it will provide tools and 10 bags of cement (to line the wells and aprons) per hand-dug well, and it will assist communities in purchasing construction material. It will also supply chemicals, such as chlorine, to disinfect the wells. The wells will be sealed to prevent contamination, and the responsibility for maintenance will be handed over to the village or town development committees (TDCs).

UNICEF's perspective involves enhancing community participation in needs assessment, planning, implementation, management, and monitoring and places emphasis on establishing affordable and appropriate technology, particularly hand-dug wells, under standardized and competent technical supervision. Other technical considerations involve training community artisans in construction and maintenance techniques. On the other hand, there are also pressures for community planning, design, construction, and supply of material and equipment to be provided by the private sector.

Key features of community ownership and management include the community's

- having legal ownership and control of the services, including formal agreements with the project agency;
- selecting the level of service it requires, can afford, and can sustain with human and financial means;
- selecting the site for water points;
- contributing real (not token) cash of between 5 and 10% to the capital cost of facilities;
- setting up a committee or board that is accountable for managing the project;
- accepting complete responsibility for operation and maintenance of the water systems, including collection, management, and safekeeping of funds and purchasing the goods and services required for maintaining the system;
- appointing its own caretakers to receive training and tools and be responsible for preventive and simple corrective maintenance; and
- being ready to undertake self-help action to assist with repairs, cleaning, and maintenance of the area around the water projects.

The premium placed on community financial obligation might create obstacles for meeting the set objectives of providing the widest rural areas with safe water. Most settlements have very small populations, and many of these, particularly in the savannah zones, cannot afford the 10 bags of cement required to construct a hand-dug well of the standard and quality proposed for a safe water supply. For example, under the Volta Region Rural Water Supply and Sanitation Project, supported by Danish International Development Assistance, communities with populations between 300 and 4 000 must first register with an amount of money (calculated as 100 GHC per person, including children and old people) to qualify for a project (Volta Region Rural Water Supply and Sanction Project n.d.). The community then pays 10% of the construction cost of the needed installation, such as the cost of a handpump for the well. Should the community "without good reason" delay payments of the monies, "the project will move on and assist the next community" that can meet the requirements.

Health could also suffer from the new policies. A 1990 UNICEF study (Government of Ghana and UNICEF 1990) found a resurgence of guinea worm epidemics in northern Ghana after GWSC instituted cost-recovery measures for pump maintenance and higher tariffs. The outbreaks occurred in villages that had potable water and where this waterborne disease had been eradicated. Because the corporation had dismantled pumps or repairs had ceased because communities had defaulted in paying arrears, the people had resorted to drinking water from ponds, dams, and streams.

Case studies

In this section, three village water systems are presented to show the effects of policy-making on water at the community level, the specific nature of the problems of water-stress systems, and how community members, particularly women and children, evolve management strategies and conservation mechanisms for their daily survival needs.

Three villages, Apollonia (Greater Accra Region), Baabianeha (Eastern Region), and Brofoyedru (Ashanti Region), were selected for detailed study. Apollonia is in the coastal savannah zone, and Baabianeha and Brofoyedru are in the tropical rain forest. Brofoyedru, located near Ghana's oldest exploited and largest gold mining industry of Obuasi, has a forest that is rapidly being depleted, whereas Baabianeha has a relatively unexploited rain forest. Brofoyedru lies in a valley of scarps, whereas Apollonia and Baabianeha are in low-lying topography.

The studies, although showing common problems and similarities in the water systems introduced, also indicate some variations in types of systems in the three communities, as well as some differences in water demand and use.

Method of study

Two principal approaches were used for the study and data collection: in-depth interviews using unstructured questionnaires in all three communities, and a focus-group discussion in Baabianeha. Thirty-four people were interviewed, about half of them women. Included in the group were one member of Parliament, two representatives of NGOs (World Vision International and UNICEF), members of the TDCs, three government officials (two from GWSC and one from the Danish Embassy in Accra), a chief (Baabianeha), members of traditional authority, a priest of a traditional religious shrine (Brofoyedru), two school teachers, two cattle ranchers, and ordinary members of the communities. The focus-group discussion in Baabianeha was conducted with a group of four women and three men. In Brofoyedru, a session was held with three young women aged between 22 and 25, including a seamstress' apprentice and an unemployed mother of two.

Apollonia

Apollonia is located 30 km between the industrial city of Tema and Accra. It is a relatively young settlement of migrants, with no distinct ethnic composition, traditional cultural practices, and sociopolitical structures. It was originally settled by people from the Ga-Adangbe ethnic group who were seeking economic activities other than fishing in nearby Kpone. Later, other settlers came, including a few Fulani men who worked at cattle rearing for owners living in Accra and Tema. The main economic activities now are crop farming and livestock rearing (cattle, sheep, and goats). Subsidiary economic activities include firewood cutting, sand winning, small-scale trading and trucking, and some artisanal production.

Traditional political authority has evolved from an age-based leadership system to a system of chieftaincy, which rotates among three families who claim original ownership of the lands. Today, the villagers elect a representative to serve on the Dangbme East District Assembly, which administers the area.

Water sources

The natural water sources for the village, with a population of 975 in 1993, are small streams, two dams, and four ponds. During the 5-month-long rainy season, there is plenty of water, but in the long dry season the streams dry up and the levels in the other sources go down drastically. In the dry season, women and children must walk to the only water source for people and animals, a dam 4 km from the village, to collect water.

The community did not take any measures to minimize the effects of the dry-season's drought, even though they knew that the small streams did not provide a constant supply of water year-round. Pollution from animals and humans competing for water in the ponds and dams was not controlled, and the water was used for bathing and swimming, even though it was also the source of drinking and cooking water. Eventually, the lack of maintenance and management led to the growth of algae and fungi, making the water unsuitable for domestic purposes. The obvious health consequence was an increase in the incidence of bilharzia.

Biogas project

In 1987, the water-supply situation began to improve in Apollonia when a biogas pilot project was implemented by the Ministry of Energy and Mines, with assistance from the governments of India and China. The project has ten 50-m^3 biogas plants that run two combustion engines for generating electricity. Twenty-one household biogas digesters, with capacities ranging from 10 to 50 m^3, have been installed. These produce approximately 190 m^3 of biogas daily for cooking and for heating water in 59 households.

BIOGAS PROJECT AND WATER — The main raw materials for generating biogas are cow dung and water. Because the water supply in the area was insufficient for the biogas project, the Ministry had GWSC construct pipelines to bring water to the community from the main water-supply system in Kpone, which supplies Accra and Tema. Before the project, the community had tried unsuccessfully to have GWSC connect it to the main system, but it could not meet the cost of piped water.

Between 1987, when it was started, and 1992, the project depended on water from dams and ponds. Because large volumes of water were required, the competition between the project and the people for the limited water supply generated considerable hostility toward a project purporting to benefit the community. At the construction stage of the biogas project, women and children were responsible for supplying water to the project from collection points that were far away, adding to an already heavy workload for women. During the operational stage, attendants, who are mostly men, were responsible for supplying water and charging the digesters.

GWSC charges the Ministry for water used in the digesters to generate electricity. Part of the tariffs paid by the users of the energy resources are remitted to GWSC to cover water costs. Individuals may also use water from the mains for domestic use, at a rate of 20 GHC for a 10 L bucket. Should the community fail to pay the rates charged per bucketful, GWSC will disconnect the supply.

The community's role in water affairs involves keeping the dams and ponds free from pollution and uncontrolled use by animals. Although some members of the community know how to keep pigs and cattle away from these traditional water sources, direct responsibility for control and supervision is still vague and spontaneous.

The community holds regular meetings with project technical officers from the Ministry, discharges the digesters, supplies cow dung for the digesters, ensures a supply of diesel fuel for the generators, and maintains the equipment and plants. They check leakages in the gas system and illegal connections to the main, especially by nearby villagers, private construction companies, and others, such as farmers along the pipelines. The Ministry's project team for the energy system is solely responsible for all important matters pertaining to the biogas project and ensures community compliance with the responsibilities assigned them. The new water-supply system and the biogas project are gradually attracting new settlers from the big cities, and there is construction of new homes, with the promise of new economic activities to uplift the community.

Baabianeha

The village of Baabianeha is 22 km from Akim Oda, a commercial and administrative centre. It has a population of 237 (1984 census). It is a farming community with no industrial or commercial activity except for the production and sale of cocoa and occasional trade in other foodstuffs when there is a surplus. The people grow maize, cassava, plantain, and, in recent years, rice for subsistence.

There have been periodic finds and sporadic small-scale mining of diamonds in the swampy forests near the village. However, these deposits have never attracted sustained and commercial investment. A nearby forest has attracted some timber exploitation, but not enough to warrant establishing a local sawmill, although Akim Oda is one of Ghana's largest wood-processing industrial towns.

Baabianeha has been settled for 12 years and is an Akan-speaking community. It has a chief and a council of elders for the indigenous political authority and, as in most Ghanaian villages, a TDC, introduced in the early post-colonial period to lead community development projects. Between 1982 and 1992, the government created a Committee for the Defence of the Revolution (CDR) in Baabianeha. The CDR and TDC work together to oversee mobilization of the community for development projects and activities, usually in cooperation with the chief. The TDC has a number of executing committees. The Water and Sanitation (WATSAN) Committee, made up of four men and three women publicly nominated in a village assembly, is responsible for managing water resources.

Water sources

Baabianeha has three water sources: streams, hand-dug wells, and boreholes fitted with handpumps.

STREAMS — The village has four streams, the Awuku, Akontan, Nkran, and Anyinasu. The villagers mostly use water from Awuku because it is closest, less than 500 m away. People who work on farms outside the village access water from the Akontan, the biggest of the streams. Palm-wine tappers and distillers of the indigenous gin, *akpeteshie*, also use water from these streams for brewing.

When the dry season is long and severe, men usually dig makeshift small wells in the dry beds of the streams to fetch the water that collects from the high water table. The water from such wells is sometimes whitish from the clay soil. Farming, especially the growing of rice, cassava, and maize, which require clearing of the vegetation, is the principal cause of the steady decrease of water, according to those interviewed. People no longer respect the old rules of keeping the banks and lands along the streams untouched. Five years ago the traditional council of chiefs in the district headquartered at Oda decreed that no one could farm within 460 m of the rivers and streams, but this regulation is not observed.

HAND-DUG WELLS — To solve the water problem, the community, which had no technical knowledge or tradition of constructing wells, sought assistance to tap the high groundwater in the Baabianeha rocks. Support came from the Primary Health

Care (PHC) Centre at the Swedru Catholic hospital. The PHC Centre provided technical personnel to locate water tables, supervise construction, and educate the community in maintenance. The centre also donated cement and provided money for the stone and sand required to construct the concrete lining of the well walls.

The community provided labour and some financial support to dig two wells 5 m in diameter. The PHC's material support went to building the well cover, the head, and the apron for preventing accidents and contamination. After the wells were constructed 7 years ago, a youth committee was put in charge of managing and cleaning them, ensuring they were covered, and weeding.

Initially, one well was dug, but the water it provided was inadequate for the whole community. The chief and the CDR agreed that the two sides of the village, demarcated by the short street, would fetch well water on alternate days until the second one was constructed. Traditional belief and ritual required that no one fetch from the well in the direction of the Awuku on Tuesdays, and it was sacrilegious for anyone to defile the water by defying this regulation.

BOREHOLES FITTED WITH HANDPUMPS — Although the wells improved the water supply, the water was still not considered as good as piped water or water from boreholes with pumps. The villagers approached UNDP for assistance to construct boreholes fitted with handpumps. They felt the boreholes would be more hygienic and a more reliable source of water than the two hand-dug wells, which have since been sealed.

In January 1994, two boreholes were drilled, and in March two pumps were fitted. The construction was done entirely by contractors brought in by UNDP and GWSC. For each borehole, the community contributed 60 000 GHC, which was raised over 2 years from adults to pay for the pumps.

After the installation of the distribution pipes, the community was entrusted with managing the new system, in coordination with the village's WATSAN Committee. The committee members are a school teacher, a nominee from the local church, a nominee of the chief, a "concerned" citizen, and three women. All members were publicly chosen or vetted and approved.

A user tariff of 100 GHC per month per adult is collected for the maintenance of the pumps. Everybody is obliged to pay, and no one is allowed to drink from the streams as an alternative to payment. To strictly enforce this the District Assembly is preparing a bylaw making it an offense to use drinking water from another source. At quarterly public meetings, the WATSAN Committee reports on the user fees. Defaulters are reported to the chief. If they persist in nonpayment, they may be reported to the District Management Committee at Oda for possible prosecution in the courts.

The technical maintenance is handled by well-trained mechanics under the District Management Committee. There are five mechanics in the area, and Baabianeha may approach any of two in nearby Awisa or Achiase when it requires services. A WATSAN Committee caretaker, trained for a day at Oda by GWSC technical personnel, handles simple routine mechanical problems, such as oiling the nuts and bolts to prevent rust or ensuring the drains are not choked. The WATSAN Committee also ensures general cleanliness of the surroundings of the pumps and makes sure children do not mishandle or damage the pumps.

The Baabianeha water project was donated by UNDP, with GWSC supervising the technical standards and implementation. However, it is the community that manages and controls the project, as it does not come under GWSC's administrative jurisdiction. The user fees collected are supposed to be kept by the village for repairs and other related expenditures.

Brofoyedru

Brofoyedru is an old village located in a valley of the Kusa scarps in the Ashanti Region in the Adansi District, about 25 km east of Obuasi and near one of the world's richest gold mines. The village has one of the oldest schools in the area and, therefore, boasts of the largest number of educated people in the area, but this has contributed to significant outward migration of young men and women.

The vegetation, once a thick rain forest, has been reduced to a secondary forest by years of exploitation for timber for the mines and by intensive clearing for cocoa and coffee cultivation. The hillsides of the scarps facing the village get lighter in vegetation as one descends. The trend is toward further denudation of the forests, gradually but steadily exposing the tops of the hills to erosion.

The location of the village exposes it to very strong, often devastating winds, which in the last decade and a half have pulled down roofs and walls of the mud houses. The destruction of several houses and the heavy migration of educated people have given the village an atmosphere of dejection, standing in stark opposition to its past glory as a bustling commercial centre from the 1930s to the 1950s. In those days, the main highway between Kumasi and Takoradi, linking Ashanti and the northern sectors of Ghana to the coast and the main harbour southwestward, passed through Brofoyedru. The village had a market and stores and warehouses for the principal colonial commercial houses. In the 1950s, a new road about 4 km away diverted traffic, and the village began to lose its commercial importance. In the 1970s the road became so badly neglected that all major transportation was routed elsewhere, and a journey of 4 h from Accra, for instance, now took more than 8 h.

Water sources

The village has three water sources: streams, the "Henderson Box" system, and boreholes fitted with handpumps.

STREAMS — A number of rivulets and streams and one river flow from the hills of the scarp and pass by Brofoyedru at various distances from the village. Many of these streams and rivulets enter the Gyimi, a sizable river that flows for more than 40 km before entering the Offin River, one of Ghana's major rivers. The Gyimi is the main source of water for the gold industry. It never dries up, although its volume is reduced during the dry Harmattan period.

For day-to-day use, three streams are of importance: the Twaano-a-antwabi ("Twa'bie"), Ahyehyentem, and Oben-ne-oben. They are all used for drinking, cooking, washing and other household purposes, as there is no industrial activity in the community. The people of Brofoyedru rely primarily on the Twa'bie for drinking and use the other two streams as a supplement when the main source is dirty. Because the water is from the hills close by, it is clear and clean and rarely dries up. This makes Brofoyedru a community with a sufficient water supply.

THE HENDERSON BOX SYSTEM — In the 1930s, a British colonial district commissioner contemplated making Brofoyedru the Adansi District administrative capital because of its commercial importance at the time. Surveys of the settlement were made for several purposes, the most important and lasting being an improved water-supply system constructed downstream on the Twa'bie. This stream is an important source of water, flowing from a high waterfall on a rocky cliff in the scarp. It then meanders through a bed of rocks and pebbles and descends gently for about 1.5 km to a spot where water is fetched by the community. About 6 years ago the Engineering Faculty of the University of Science and Technology, Kumasi, undertook feasibility studies on the waterfall with the objective of constructing a small hydroelectric project on the stream.

At the foot of the hill near the village, a filtration system was constructed in the 1930s by the European engineers from the mining and administrative town of Obuasi. The respondents interviewed refer to the system as the Henderson Box, although this name is yet to be confirmed from the literature. The system comprises an open reservoir with a capacity of about 273 m^3 and a metal sieve to block debris from weeds and trees as the stream empties into the reservoir. Another metal sieve cleans the water as it flows through a 250-mm diameter steel pipe into a smaller (approximately 23-m^3) closed reservoir with a large, removable slab on top. Four 50-mm diameter galvanized steel outlets, with steel filters inside the wall, serve as the pipes for collection.

This system produced very clean water. However, the fate of the system is a classic case of mismanagement. Until the 1960s, the responsibility for technical maintenance lay with the sanitary inspectors of the District Council. The community cleared the weeds and periodically desilted the reservoirs, under the authority of the TDC. Desilting, which takes 2–3 d to complete, was done by adult men as imperative communal work. No one in the village knew how to maintain the system, nor did the TDC have the technical know-how to rehabilitate the system. The last cleaning exercise was done in 1992. In recent years the concrete structures sank in because the foundations and walls cracked from lack of maintenance. Water seeped out, weakening the base and the ground around the reservoirs. GWSC received complaints but made no practical response.

The fact that the system still maintains traces of life more than half a century later makes it worth investigating and possibly reviving. The stream continues to flow into the reservoir and by all indications would continue to provide water if the system is rehabilitated. The community respects the TDC regulation protecting the forests at the headwaters in the hills and along the stream from destruction by any activity, thereby preserving the integrity of the stream. Even in its broken-down state, the people still consider the Henderson Box the best source of drinking water.

BOREHOLES FITTED WITH HANDPUMPS — As the reservoir became harder to maintain, the TDC requested an alternative water system for Brofoyedru. GWSC constructed five boreholes fitted with handpumps in the village in 1988 and 1989. Three years after the system was brought in, GWSC sent a bill for 500 000 GHC for the community's arrears for water use. The people contend there was no notification of charges when the projects were introduced, and they expected none. For more than 1 year, four of the pumps have been shut down, disconnected by GWSC for nonpayment of bills. Now GWSC charges 200 GHC per adult per month as water rates. The community claims it cannot pay the arrears and feels no urgency at all about settling the bills and getting the pumps reopened. The people claim that the water from the boreholes does not taste as good as the water they are used to from the reservoir. It is also argued that the tariffs are unjustifiable because, after the installation of the pumps, GWSC did not contribute to the production of the water from the boreholes by providing chemical purification or supplying energy to power the system.

Discussion

Apollonia is in transition toward becoming part of Accra or Tema. Its water-supply system is an indirect outcome of a biogas-energy development project in which the community did not wholly participate. If the Ministry were to withdraw from the project, it is not clear whether the community would continue maintaining it, as it does not own or control any aspect of the water pipelines.

The villages of Baabianeha and Brofoyedru both use boreholes with pump-fitted wells. Baabianeha has a system of local ownership and control that has been successful, in part due to a UNDP donation that reduced economic pressure. The monthly user fee of 100 GHC per adult is yet to prove a problem, but the test will come in 1996, when, during the March–August lean season, people have little income. In Brofoyedru, the investment in new wells means that the traditional technology, the Henderson Box reservoir, has been abandoned. The new borehole system stands unused, an "alien" edifice, because GWSC did not inform the people of their financial obligations when the projects were introduced and did not consult with them to determine their actual capability to meet those financial demands.

Local management issues

In both Baabianeha and Brofoyedru, access to the technicians for maintenance appears to be a problem, as there are only five in the whole district. It would be feasible for a community volunteer to be trained to handle more complex repairs (beyond basic corrective and preventive exercises) to avoid and repair breakdowns.

The problem of conservation for the Baabianeha community lies in how to discourage the continued exploitation and denudation of the vegetation around the streams. This complicated question involves traditional authority in the district and the state, which the people can least influence. Baabianeha's WATSAN Committee adequately represents the village's traditional social order and has the potential to manage and take care of the village's water issues, although results would be improved if there were more women on it.

Upgrading traditional water systems

It seems that abandoning and sealing the two hand-dug wells in Baabianeha was not a good idea. There were shortages of water in the wells and excessive levels of contamination, but it might have been preferable to disinfect, rehabilitate, and reinforce them and strengthen the structures that prevent accidents and contamination. The community would then have had additional sources without significant new investment.

Brofoyedru exemplifies a case GWSC provided an alternative water system but did not take into account the existence of other sources that needed improvement and did not consult the community about tariffs. It is also a good example of how a community, pressed by financial obligations it cannot meet, will revert to traditional sources of water, despite the health risks. Getting water from traditional sources may be acceptable in communities where the villagers have a strong attachment to the stream and are disinterested in hand-dug wells or even boreholes with pumps.

The village also demonstrates the problems of poor education in relation to the quality of the water from the new system. The insistence that the taste of the new water is not up to the quality of the old source and that the borehole water stains enamel and aluminum plates and utensils shows an absence of any serious education about the project. This community's problems with the borehole system may have been minimized or avoided had the suitability of the half-century-old filtration system on the local stream been investigated.

It is obvious that the lack of a clear policy and implementation framework often leads to neglect of water systems, creating stress that affects the kind of water available for rural communities. The case studies demonstrate that, with respect to traditional water resources like streams, rivers, ponds, and springs, local management of ideas, institutions, and practice has been absent or weak. There have been spontaneous efforts to keep the sources clean, but without any studied regularity and without the input of women in the relevant decision-making. Folk songs, stories, myths, and traditional practices could be used to educate the communities in water use, conservation, development, and management and to encourage positive attitudes toward natural water sources.

Alternative technologies

Can alternative water-supply systems that are affordable, acceptable, and safe be developed? Indeed, a number of other technologies could play a greater role in Ghana. Training would, for example, make feasible the introduction of spring boxes, rainwater harvesting, infiltration galleries, and household water filters. Although these technologies are appropriately modest in cost, it must be recognized that they are generally not alternatives, but important supplements.

Despite the success of rooftop rainwater catchment elsewhere, few villages in Ghana today use this technique. Houses in the smallest communities, mostly in the savannah, have thatched roofs. Moreover, annual rainfall in these areas is so low that harvested water could only be significantly useful if the technology is elaborate enough to combine a number of households and to provide sanitary communal cisterns for long-term storage. The opportunity for rooftop catchment is not much better even in those parts of Ghana where rainfall is higher and where aluminum roofing is common. The quality of the aluminum roofs is so poor that rainwater catchment might require reconstructing of the houses. Overall, then, it seems rooftop catchment must await a long-term linkage with reforms in housing design and construction.

For those communities that do not yet have safe drinking-water sources or cannot afford them, a UNICEF proposal provides a realistic and economically accessible alternative:

- Filters and chemical agents should be distributed to communities without wells or safe water.
- Cloth filters should be distributed to people relying on stagnant water, particularly on farms.
- More efficient and affordable filters using sand, charcoal, and gravel should be constructed in homes in such areas.

To strengthen this approach, it will be necessary to study the socio-economic context within which an alternative water system could be implemented. The nature of the household and of the social relationships, particularly gender and how it affects community relationships and the participation process, needs to be considered. This will ensure the active participation by different actors at different levels of decision-making in the implementation of water systems.

To improve water use and management, there is a need for a systematic education program using various mass media, interpersonal communication techniques, and social occasions (such as festivals, market days, and literacy classes). However, the educational program will only succeed if other factors such as the water system's cost and tariffs are not so great as to militate against receptivity and acceptance.

Conclusions

This study sought to examine the interrelationships between government policy on water, the use of natural and alternative sources of water, the existence of water stress and traditional coping strategies, the involvement of different actors at the community level, and the management of water systems.

It is clear from the study that government policy on water, within the framework of the SAPs, has overlooked the importance to women and children of good sources of drinking water. There has been a shift from providing water as a basic need and social right for rural people to putting pressure on their communities to provide some of the initial financial capital as a prerequisite for obtaining water. In a situation of diminishing incomes, increasing numbers of households headed by women (where there is usually dependence on one income), and continued impoverishment of a majority of the rural people, this requirement makes it almost impossible for communities to get potable water. This has resulted in continued dependence on stressed systems, which has negative implications for the health of women and children. The traditional knowledge of users is usually not taken into account in defining the roles of people in water management, resulting in a lack of commitment among rural communities to comply with a framework designed from the top. The integration of indigenous knowledge into the management of water systems would empower communities to take their lives into their own hands.

The government should review its policies toward social-health issues like water. If health is to be given priority, water supplies to poor communities should be subsidized.

In GWSC's concept of community ownership and management, communities are expected to contribute to the establishment and maintenance of improved water-supply systems. However, the proposition that communities pay 5–10% of the cost of the construction material, particularly for handpumps, should be reviewed, especially for communities with populations below 500. In the dry savannah zone, there are many communities that may never collectively earn enough to spare 50 000 GHC a year.

The role for communities is by no means limited to finance and maintenance. Communities must be involved, from the start, in decisions about which water systems they want, what they can afford, and where the systems should be installed. Such involvement can only improve decisions about the introduction of technology that is affordable and accessible, both in economic terms and in terms of the acquisition of technical maintenance skills.

When localities receive financial and technical support from institutions like UNICEF, their involvement in the various stages of decision-making — assessment, planning, implementation, management, and monitoring — enables them to evolve a good management strategy that can sustain the alternative water system. This approach respects the fact that rural people are aware of their own needs and have indigenous knowledge of managing water for their own use. Participation leads to better gender relations and the improved health of household and community members.

References

Government of Ghana; UNICEF (United Nations International Children's Emergency Fund). 1990. Children and women of Ghana: a situation analysis. Government of Ghana and UNICEF, Accra, Ghana.

——— n.d. Ghana master plan of operations and programme plans, 1991–1995. Government of Ghana and UNICEF, Accra, Ghana.

GWSC (Ghana Water and Sewerage Corporation). 1993. Survey 1991. GWSC, Accra, Ghana.

Parliamentary Committee on Works and Housing. 1994. Report to Parliament. Accra, Ghana.

Volta Region Rural Water Supply and Sanitation Project. n.d. Project information booklet.

World Bank. 1991. Water supply and sanitation in Africa: laying the foundation for the 1990s. Proceedings, All-Africa Rural Water Supply and Sanitation Workshop and Water Supply and Sanitation Conference. Vols. 1 and 2. World Bank, Washington, DC.

Water Management, Use, and Conflict in Small-scale Irrigation: The Case of Rombo in the Kenya Maasailand

Hartmut Krugmann
Senior Program Specialist, IDRC, Nairobi, Kenya

Introduction

Population growth and other factors put increasing pressure on freshwater resources in Africa. Among the different water end-uses in Africa, the agricultural sector accounts for the lion's share. It is sensible, therefore, to examine how water is used in existing irrigation systems, with a view to efficiency management. In sub-Saharan Africa, large-scale irrigation systems have not fared well because of development costs, management problems, and inefficient water use. It is generally agreed that small-scale schemes building on local knowledge and institutions hold greater promise (Rached 1994), and it is important to focus attention on such schemes and accumulated experience.

This paper presents preliminary results of field research into smallholder irrigation activities and other water uses in Rombo Location, Loitokitok Division of Kajiado District, Kenya. The findings are based on a series of in-depth interviews with different resource users and other key informants. The research is an integral part of a larger study assessing processes of land degradation in the area. This paper examines ecological, economic, social, institutional, cultural and political aspects of water access, use and management, primarily in irrigation, and highlights factors contributing to local water stress and conflict as well as possible avenues for conflict resolution and improved management.

Characteristics of the study area

Rombo Location is in the southeastern corner of Kenyan Maasailand, close to the foothills of Mount Kilimanjaro and bordering Tanzania. It is mostly semi-arid: rainfall is highly seasonal and variable, and water becomes a critical constraint during the dry season and even more so during longer spells of drought.

Within Loitokitok Division, Rombo Location is the lowest lying subregion, at an altitude mostly just above 1 000 m. Mean annual rainfall ranges from around

400 mm in the plains of the subregion to 1 000 mm in the foothills of Mount Kilimanjaro. Approximately 80% of Loitokitok Division falls into the category of "low-potential" land and is predominantly used for extensive livestock husbandry by the indigenous Maasai people.

About 10% of the land in the division is "medium-potential" land, which is used for livestock-keeping, with cultivation predominant in the better watered areas. Finally, the remainder, another 10% or so, is "high-potential" land. Here, rainfed cultivation predominates. In addition, irrigated cultivation is practiced in a number of pockets in the plains where springs and swamps supply water even in the dry season.

Land tenure

The above classification of land is mirrored in existing land-tenure forms. The low-potential land tends to be held under group title (land divided up in so-called group ranches). Most of the medium-potential areas are divided into large, individually owned ranches ranging in size from 100 to more than 1 000 ha. Finally, small title plots (mostly less than 4 ha) can be found throughout the high-potential areas. These tenure forms were chosen when all the land of Kajiado District was adjudicated around 1970. However, there is considerable pressure for group ranches to subdivide into individual, private smallholdings.

Fieldwork for the present study focused on two sublocations of Rombo Location: Rombo and Njukini sublocations. This area comprises the Rombo Group Ranch, several individual ranches, and farmed land (rainfed or irrigated) held by thousands of smallholders under individual, private title.

History of socioeconomic change

To understand the socioeconomic and environmental realities in the Rombo area, it is useful to briefly review the recent history of land use and demographic changes. Before the onset of colonization in 1895, there was only one dominant land-use system practiced by the Maasai, namely, nomadic pastoralism. This resource-use system relied on the plains for wet-season pasture and on elevated areas and places endowed with permanent water sources for dry-season grazing. The system was well adapted to the local ecology and hence sustainable.

The colonial rulers expropriated large stretches of land previously controlled by the Maasai. Two treaties in 1904 and 1911 (the so-called Maasai Moves) confined the Maasai to a southern reserve, essentially today's Kenyan Maasailand (districts of Kajiado, Narok, and Trans-Mara). This is an area that is likely to have always been an interface between Maasai pastoralists and agriculturalists, with the existence of Ilcurran communities and some degree of intermarriage with adjacent ethnic groups.

Although the southern Maasai Reserve was sealed from outside influence throughout much of the colonial period, in-migration of other ethnic groups did occur. In particular, colonial expropriation of fertile lands in central Kenya created vast numbers of squatter Kikuyu communities, many migrating in search of better livelihoods. Some succeeded in settling within the Maasai Reserve. These early Kikuyu settlers further introduced and spread agriculture among the Maasai.

In addition, colonial policies of "pacification" and "sedentarization" of pastoralists, promotion of agriculture, creation of nature reserves, penetration of market forces, and population growth undermined the viability of the traditional pastoral system. What used to be a unified, extensive land-use system started to fragment into different forms of competing land use, including rainfed and irrigated agriculture, as well as wildlife tourism in the newly established national reserves and parks. As a rule, the land with highest potential and the most water was progressively made inaccessible to the pastoral livestock economy.

The demise of the colonial regime did not significantly change state policies. In fact, the sedentarization of pastoralists, the establishment of national parks, the promotion of agriculture, and the "commoditization" of the local economy proceeded at an even faster pace. Privatization of land tenure, in particular, reinforced the market principle and gave rise to widespread land transfers and sales. At the same time, the opening of the Maasai Reserve after independence allowed a rapidly increasing stream of migrants, mainly cultivators, into the area. Many bought high-potential land at bargain prices from the Maasai, who, at least initially, did not have an appreciation for land as a precious market commodity. In addition, large numbers of migrants, some from neighbouring Tanzania, have entered into informal short-term land-leasing or share-cropping arrangements with the Maasai. Migrant populations throughout Maasailand have grown very fast and are approaching the size of the indigenous Maasai population.

These processes have stratified Maasai society and have led to growing inequality of resource access and wealth. At the same time, traditional forms of production, reproduction, and social organization have been fast eroding and are evolving into new adaptations in the context of rapidly changing economic, political, and ecological realities.

Water sources, uses, and constraints

Water has always played a crucial role in the local land-use systems of Kajiado District. The traditional Maasai pastoral economy developed an elaborate system of moving herds and people to have access to critical dry-season water points. Later, after independence and especially with land adjudication and individual titling of high-potential and well-watered lands around 1970, these permanent water sources became increasingly inaccessible to livestock and pastoral people. Access to permanent water was largely lost as the extensive livestock economy became encapsulated through the establishment of national parks and the spread of agriculture on freehold land and in recent years even on group ranch land.

This encapsulation is very evident in the Rombo–Loitokitok area. Tsavo National Park was created on the eastern side; Amboseli National Park, on the western side. Further, high-potential land on the slopes of Mount Kilimanjaro (rainfed) and around permanent water springs and swamps in the plains (irrigation) was appropriated for agriculture. These developments removed critical water sources from the livestock system and put it under considerable water stress.

Today, the main permanent surface-water sources in the Rombo–Loitokitok region are springs and swamps in the plains at the foot of Mount Kilimanjaro. These springs and swamps work like artesian wells and are pressurized and recharged by Mount Kilimanjaro water. The two largest springs in the Rombo area are Kisioki and Olchoro. In most cases, springs discharge into existing river beds, like the Rombo River, which otherwise would dry out during the dry season.

The most important engineered water structure in the division is a 100-km pipeline taking Mount Kilimanjaro water from Noolturesh (near Loitokitok) to areas less endowed with water. There are few engineered groundwater-harvesting structures like boreholes and shallow wells in the area, and the untapped potential for groundwater supply is large.

Domestic water use is also put under strain by growing water scarcity (and diminishing quality). Women, in particular, are affected because fetching water is one of their numerous tasks. On the Rombo Group Ranch, this can take up to 3 h every other day, depending on distance to the water source and size of the family.

Water quality is a serious issue in the Rombo area. Irrigated horticulture involves copious use of pesticides. Toxic chemicals are flushed into the river during heavy rainfalls. Furthermore, people use the river to wash clothes, cars, or themselves. At the same time, water downstream from irrigation schemes or other polluting activities is routinely used as drinking water and for other domestic purposes. Water pollution can also have harmful effects on income-earning activities (notably honey production, because bee-keeping requires clean water).

Irrigation

Irrigation activities in the Loitokitok area were initiated in the 1950s. Early irrigated cultivation involved just a few families and primarily met subsistence needs. It was only in the 1970s, after land adjudication and privatization, that irrigated land and numbers of irrigators expanded significantly and cultivation was directed toward commercial purposes.

Today, Loitokitok Division is the centre of irrigated cultivation in Kajiado District: 11 out of 16 small-scale, farmer-managed, group irrigation schemes in the district are found in this division. Table 1 provides some information on irrigated areas and numbers of irrigators for these irrigation schemes.

The total estimated area under irrigation in 1993 was 1 256 ha, which was farmed by 3 219 families (DIU 1993). The mainstay of farming activity is horticulture, and the main horticultural crops are onions, tomatoes, and a variety of so-called Asiatic vegetables, such as chilies, karela, okra, and brinjals.

Existing irrigation schemes in Loitokitok mostly involve families who farm individual smallholdings. However, to a significant extent, irrigation is also being

Table 1. Irrigation schemes in Loitokitok Division.

	1988		1994			
Scheme	Members [a] (*n*)	IA_1 [b] (ha)	Members [a] (*n*)	IA_2 (ha)	IA_1 (ha)	GA (ha)
Rombo	734 [c]	395	1 125	539	380 [d]	979 [d]
Kimana–Tikondo	700	320	717	162	119	758
Namelok	375	250	468	864	417	970
Isinet	120	73	149	115	66	132
Empiron	—	—	—	209	43	295
Inkisanjani	220	122	—	—	—	—
Illasit–Olkaria–Entarara	93	73	—	—	—	—
Elangata–Enkima–Oloorika	160	65	—	—	—	—

Source: 1988 statistics are from Kimani (1988, table 1.1); 1994 statistics, from Ewaso (1994).

Note: GA, gross area; IA_1, irrigated area; IA_2, irrigable area;

[a] Number of registered irrigation farmers.

[b] No distinction is made between irrigated area and irrigable area in Kimani (1988). Possibly, in most cases, pressure on irrigation water was not yet high enough for irrigated and irrigable areas to be different.

[c] Includes Upper Rombo and Lower Rombo (Maili Tatu).

[d] This figure is taken from DIU (1993), as it was not available in Ewaso (1994).

practiced on group ranch land. Some schemes are located entirely on group ranch land, and others extend onto group ranch land. For example, the Inkisanjani and Oloorika schemes are located on the Kuku Group Ranch (see Table 1), and three of the 13 furrows of the Rombo irrigation scheme (RIS) — Kanan, Karanja, and Olchoro — extend onto the Rombo Group Ranch (see Table 2). In these cases, "landlords" do not own their land, but they have received permission (formal or informal) from the group ranch authorities to cultivate.

Case study: Rombo irrigation scheme

RIS is one of 11 schemes in Loitokitok Division. It is the largest (in terms of the number of irrigators) and one of the oldest in the division. It consists of two subschemes, the Upper Rombo and the Lower Rombo (Maili Tatu–Ntere Kesi), both supplied by the two main springs in the area, the Kisioki Spring and the Olchoro Spring, which discharge into the Rombo River.

In comparison with the other irrigation areas, Rombo has a warmer climate because of its lower elevation and more rainfall because of its proximity to Mount Kilimanjaro. More than 80% of crops grown locally under irrigation are horticultural produce: onions and tomatoes for national markets and Asiatic vegetables for export overseas, mainly to the United Kingdom. Although onions and tomatoes are common horticultural crops in the region, Asiatic vegetables are particular to Rombo. Rombo is the principal centre in the division for the production of these Asiatic vegetables, which gives it an international trade link.

Evolution of the scheme

The first irrigation activities in the Rombo area started around 1950 (PIU 1986). They were initiated by maverick Maasai who had gained farming experience with Wachagga (Tanzanians) or Indians (Kenyan Asians) (Kimani 1988). Early irrigated cultivation remained limited to a few families and served subsistence purposes. The idea of expanding and improving irrigation activities was first raised by the local Roman Catholic mission in 1965. Subsequently, under a policy of land adjudication, land was demarcated into individually owned blocks of approximately 12 ha each.

Over time, more people were attracted to the area by the potential for more intensive (irrigated) agriculture. Most of those who settled (Kenyans) worked as tenants on land owned by the indigenous Maasai. Some were able to purchase land, often at bargain prices, from the Maasai. Increasing in-migration and the establishment of irrigators put pressure on available water resources and caused serious problems of water access.

This led to the formation of the first irrigation committee, which consisted of elders. At the same time, water distributors were appointed and temporary intake canal structures were constructed. The irrigation committee attempted to restrict the influx of new water users by introducing an entrance fee. In 1976, for the first time, bylaws regulating access and distribution of water were drafted and went into force. Later on, the two irrigation sections (Rombo and Maili Tatu) formed their own committees. In addition, most individual furrows formed their own committees and appointed their own water distributors. In 1985, the two subscheme organizations became part of an overall committee, the Rombo Irrigation Committee, whose function was to coordinate water allocation within the entire RIS.

Individual furrow committees, as well as the overall committee, were supposed to be elected annually. However, because of internal squabbles, this did not always happen. Furthermore, the election process was not always transparent. Particular local or outside interests sometimes significantly influenced the outcome of elections. The elections in November 1993 were very politicized, and only indigenous influential Maasai were elected as committee members, irrespective of their knowledge of irrigated cultivation and collective irrigation management (Ewaso 1994). Notwithstanding this turn of events, farmers' organizations in RIS have been classified as strong in comparison with other schemes in the division (DIU 1993, table 2.1).

Irrigation technology, layout, and water distribution

Like most other irrigation schemes in Loitokitok Division, RIS was developed and is managed entirely by the local farmers without outside assistance.

Technology

The technology of water diversion and conveyance is simple and rudimentary. Water is taken from Rombo River by means of simple check structures erected across the river bed. These structures are made of tree branches, stones, and soil. During floods in the rainy season, intake structures are often damaged or washed away entirely. Reconstruction or repair requires considerable labour and time.

Check structures in the river bed raise the level of water up to intake levels of conveying canals. Water then flows down the canals and subcanals, driven by gravity, until it reaches the farmer's field. There is no provision for on-river storage (like dams). Therefore, at any given time, the availability of irrigation water depends entirely on the yield of the springs that discharge into the river.

Like the intake structures, irrigation canals and most of the divisions into subfurrows are temporary (made out of soil and other local materials). Only three of several dozen divisions are permanent, and none of the furrows or subfurrows are lined. There are no permanent flow-control devices like check gates, drops, or sluices. Furthermore, no more than makeshift structures have been put in place where furrows cross roads, livestock paths, or other obstacles. Water losses during conveyance (seepage from canal bottoms, spillage from breached canal banks, leakage from gully and road crossings, etc.) are correspondingly high.

The type of irrigation practiced throughout the Rombo area is flood irrigation using field basins, a technology that is simple but does not use water efficiently. Water is conveyed to individual plots by means of head and in-field ditches, which empty into field basins. There are two basic types of basin structures: ridged basins, used primarily for tomato-growing, and flat basins, used mostly in the cultivation of onions.

Layout

There are 13 canals originating from Rombo River or separate smaller springs or swamps. Table 2 provides an overview of the main furrows, including the total farm area covered by each of them and the size of the membership of farmers' organizations (individual furrow committees), as well as their ethnic composition. Furrows are listed in sequence starting upstream (Kisioki Spring) and going downstream (Kanan is the spring that is farthest downstream).

Table 2 also shows for each main furrow whether farmers are already using all available water or whether there is room for expanding land under irrigation using current technology. The table also indicates whether, within the latter category, a possible expansion would be marginal or significant. It is evident from Table 2 that only 2 of 13 furrows would allow significant further expansion of land under irrigation if there is no change in irrigation methods. All other furrows would require upgrading of irrigation structures and more efficient in-field water use to increase the scale of total water services for irrigation.

The readily apparent overall water constraint for irrigated cultivation in the Rombo area is underscored by the results of a recent water-availability assessment study carried out for all irrigation schemes in Kajiado District. This study is summarized in DIU (1993, pp. 18–21). The study concludes that for RIS (and for most other irrigation schemes in Loitokitok Division) "a serious water shortage already does occur at least once in five years for present water abstracters — no consideration for further scheme development should be given."

Even though this conclusion is based on a number of questionable assumptions and rather shaky streamflow data, the general thrust of the argument is likely to be valid.

Table 2. Rombo irrigation scheme.

Furrow [a]	Members [b] (n)	GA [c] (ha)	Water availability	Ethnic composition
Upper Rombo				
Kisioki	300	240	Sat.	Ka–Ki–M–C
Loishiro	60	48	NS/M	Ka–Ki–M–C
Chacha	27	20	Sat.	Ki–M (50 : 50)
Njeru	40	50	NS/M	Ki–M
Ole Kidoku	66	54	Sat.	Ka–M–Luo–C
Kisopia	90	72	Sat.	Ka–Ki–M
Nduma	5	4	Sat.	Ki–M
Nesere	22	15	Sat.	Ki–Ka–M–Kal
Lower Rombo				
Olchoro	30	24	NS/M	Ka–C–Luo–M
Embolie	94	45	NS/M	C–Luo–Ki–Ka–M
Karanja	60	60	NS/S	Ki–C–Ka–M
Essosion	196	185	Sat.	Ki–M
Kanan	135	162	NS/S	C–M–Ki
Total	1 125	979		

Source: Mr Otieno (Horticultural Officer) and Mr Njau (Irrigation Officer), Ministry of Agriculture, Rombo (personal communication, 1995).

Note: C, Chagga; Ka, Kamba; Kal, Kalenjin; Ki, Kikuyu; M, Maasai; NS/M, nearly saturated (marginal expansion possible; NS/S, not saturated (significant expansion possible); Sat., saturated (all available water used).

[a] List starts upstream and goes downstream.

[b] Number of registered furrow members.

[c] Gross area of farmland linked to furrow.

Water distribution

Irrigation activities in all individual furrow systems proceed continuously day and night. Use of irrigation water rotates among farmers within given subsections of any particular furrow system. Water-distribution schedules are established by the furrow irrigation committee and overseen by the water distributor of the sub-section in question.

The main indicators for water availability and use for irrigation are the duration of the period during which any particular farmer within a furrow system has continuous access to water and the interval between any two water allocations to the farmer. As irrigation schemes or individual furrow systems expand and as the number of participating farmers increases and water demand rises, the duration of the water share tends to decrease and the irrigation interval tends to increase.

Farmers' organizations

Each furrow has its own farmers' organization. Farmers who want to irrigate must register with the farmers' organization of their particular furrow system to be entitled to water allocations. Registration is open to both landlords and tenants. Although entrance fees have risen over time, to 500 KES (in 1996, 42 Kenya shillings [KES] = 1 United States dollar [USD]) since around 1983, it is doubtful that they have ever been effective in deterring newcomer irrigators from joining farmers' associations.

Farmers' associations form their own irrigation committees, primarily at the furrow level. Furrow or spring committees typically have the following responsibilities:

- distribution of water to the farmers;
- reallocation of water shares among farmers in situations where some farmers do not want to use the water for a limited period (such as a season or a year) or decide to leave the farmers' association for good;
- mobilization of farmers for voluntary work to reconstruct, repair, or maintain irrigation structures;
- organization of communal meetings, such as regular (often weekly) meetings to determine and communicate the timing and rotation of water allocations among farmers;
- enforcement of bylaws and mediation of conflicts between irrigating farmers; and
- collection and banking of funds contributed by farmers (or, in very few cases to date, outside agencies) and registration fees from newly registered members.

The implementation of some of these tasks, especially water distribution and reallocation and mediation of conflicts, is normally left to water distributors, who are elected by the farmers or appointed by the committee. Furrow systems may have one or several water distributors, each in charge of a subsection of manageable size.

Where operational issues cannot be resolved by a water distributor, they are passed to the committee. Water distributors are usually not salaried but work as volunteers. However, their pivotal position in the day-to-day management of water resources and irrigation schedules brings them benefits through "informal" gifts, in cash or in kind, from farmers wishing to keep good relations or to gain special favours.

Water stress and institutional responses

Expansion of the irrigation scheme

In RIS, as in most other irrigation schemes in Loitokitok Division, the pressure on available irrigation-water resources has been increasing over the years. Water-use pressure and lack of access prompted the establishment of the first irrigation committee in the 1970s, but despite perceived water scarcity, new irrigator farmers have continued to arrive and settle in Rombo. Table 3 lists the number of participating farmers over the last 10 years. The membership of registered irrigator farmers in the whole of Rombo has nearly doubled over this period, reaching 1 125 in 1994.

Irrigation activities in Lower Rombo developed later than in Upper Rombo. Most of the recent growth in membership has taken place in Lower Rombo, in particular the Kanan, Karanja, and Embolie furrows. Nevertheless, even the membership in the oldest of all furrows, Kisioki (in Upper Rombo), increased by as much as 50%. Clearly, existing farmers' associations and irrigation committees have been unable to restrict new entrants through fees or other means. Interference of local power structures with irrigation business, nepotism among local influential groups, and lack of transparency and accountability in irrigation management are likely to have contributed to this situation.

What impact has scheme expansion had on water allocations to individual farmers? There has been a general trend in virtually all of the furrows toward longer periods of rotation of water shares. In the mid-1980s, intervals averaged approximately 1 week (PIU 1986). Today, intervals can be as long as 2 weeks, and in at least one case (Essosian Furrow) the interval in 1994 degenerated into something like 30 d. Adjustments in the duration of water application toward shorter water allocations have also occurred. For example, in the not-yet-saturated Kanan Furrow (Lower Rombo), water shares have been reduced from 7 h (every week) 3 years ago to 6 h (every 2 weeks) now (N.O. Kilumet, Rombo, personal communication, 1995). Share-splitting practices have

Table 3. Expansion of Rombo irrigation scheme membership.

Furrow	Membership 1994	Membership 1989	Membership 1994	Gross area (ha)
Upper Rombo				
Kisioki	200	250	300	240
Loishiro	30	47	60	48
Chacha	10	15	27	20
Njeru	13	27	40	50
Ole Kidoku	35	41	66	54
Kisopia	55	71	90	72
Nduma	3	5	5	4
Nesere	5	18	22	15
Subtotal	360	474	600	503
Lower Rombo				
Olchoro	12	18	30	24
Embolie	18	24	94	45
Karanja	6	30	60	60
Essosian	150	173	196	185
Kanan	62	71	135	162
Subtotal	248	316	525	476
Total	608	790	1 125	979

Source: Mr Otieno (Horticultural Officer) and Mr Njau (Irrigation Officer), Ministry of Agriculture (personal communication, 1995).

even further. However, because of transit times for water to reach remote plots and for other reasons, durations have not been reduced to less than 2 h.

Another effect of an expanding irrigation system (and reflection of water stress) is reduced water-application rates. Longer transit distances mean larger water losses and hence reduced water flow.

Markets in irrigation-water shares

To some extent, reductions in duration have gone hand in hand with trading and reallocation of water shares among furrow members. Some irrigators prefer to give their share temporarily to someone else rather than use it themselves, particularly if they lack the resources for the necessary farming inputs or feel that available water is insufficient to ensure adequate crop output. In situations where farmers are able to accumulate water shares, water reallocation is also more frequent. Water reallocation is a responsibility of the irrigation committee. However, in practice, water shares are often reallocated ad hoc, with the assistance of the local water distributor. This provides greater flexibility but is possible only when the rearrangement takes place within the distributor's section of responsibility.

In some furrows where water scarcity is acute, an informal market for water shares has evolved. Shares are traded unofficially for prices determined by local demand and supply, even though water itself has always been "free." For example, until recently, farmers belonging to the Essosian Canal sold and bought water shares for 500 KES. That local market in water shares was, however, brought to a close with a committee decision in August 1994 to reallocate shares on a more equitable basis (one share per farmer, regardless of the size of the plot).

Trading in water shares can provide farmers with more flexible choices (using, buying, or selling shares) and thus can lead to greater efficiency in water use. However, none of my interlocutors (committee member, distributor, or irrigator member) seemed to be in favour of formalizing such a water market. The reasons differed: the committee member saw the perceived risk of reduced control from the vantage point of the committee; the water distributor may have been mindful of the "fringe benefits" he might be losing; and the farmer member invoked equity considerations, that is, preventing rich farmers from buying up a disproportionate share of water allocations to the detriment of the poorer members.

Effects on agricultural productivity and size of irrigated area

Excessively short durations, long intervals, and low rates of water application in irrigation pose clear risks to productivity. Horticultural crops do not thrive unless water is applied in suitable doses. Tomatoes, in particular, require water every 7–10 d at a minimum. Where irrigation intervals exceed this limit, arrangements may be made between farmers (facilitated by the water distributor), usually in the same furrow section, to split water shares into two half shares of half the duration received at half the interval. This helps the farmers avoid productivity losses.

Although splitting shares may address the problem of excessively long intervals, allocations of 2 or even 3 h may not be sufficient to irrigate even 1 ha, especially when flow rates are lower. Hence, farmers often resort to irrigating only a fraction of their irrigable land. In Rombo, such a reduction in irrigated area per farmer appears to have outweighed increases in numbers of irrigators. Indeed, Table 4 shows that the total area of land under irrigation in Rombo decreased from 440 ha in 1986 to 380 ha in 1993.

Table 4. Recent evolution of irrigation in Rombo.

Category	1986	1988	1993
Membership	670 [a]	734	1 066
Irrigated area (ha)	440	395	380
Irrigable area (ha)	539	539	539
Gross area (ha)	979	979	979

Source: 1986 statistics are from PIU (1986); 1988 statistics, from Kimani (1988); 1993 statistics, from DIU (1993).
[a] Estimated on the basis of the available figure for 1984.

No great efforts appear to have been made to upgrade temporary irrigation structures to reduce copious water losses. The only major upgraded (permanent) structure to date is a gully crossing on Essosian Canal. Likewise, efficiency of in-field water use does not seem to have improved significantly.

Water conflicts and their resolution

Conflict among irrigators

Irrigated farming in Rombo is characterized by a significant level of ongoing conflict between competing water uses and users. Conflicts arise at different levels: within particular furrow systems (within or between subsections); between different furrows; and between irrigation and other uses of water. The most common cause of conflict within and across furrows is water stealing. Because water is scarce and profits depend on having access to sufficient amounts at a particular time, farmers may try to extend the duration of their water allocation at the expense of the next one in the queue.

Moreover, some farmers may tinker (often overnight) with the intake of other nearby farmers to get some water even when it is not their turn; this is easily done (and not always detected) because all intakes and furrows are simple, temporary, makeshift structures. Anger and conflicts can also be caused by reduced flows or complete lack of incoming water at the time when water has been officially allocated. This may occur quite frequently for irrigators toward the end of a furrow. This problem is often contributed to by farmers at the beginning of the furrow when they take advantage of their privileged location to clandestinely divert some of the water passing by. Other conflicts arise when members do not participate in mandatory communal work on irrigation canals or intakes.

The first level of mediation for farmers' quarrels is usually the water distributor. If the distributor cannot resolve the conflict, it comes before the irrigation committee for resolution. Repeat offenders are fined, or in serious cases they may even lose their water allocation. Fines depend on the particular kind of offense but are generally low (500 KES). The level of the fines does not represent a significant deterrent for farmers.

At the interfurrow level, there is constant competition and hence a latent potential for open conflict. This is especially so during the dry season and between neighbouring furrows receiving water from the same spring. When the quantity of water flowing into their own main intake is considered insufficient, farmers regularly resort to collective sabotage by "adjusting" (overnight) the intake structure of another furrow. There may be back-and-forth mutual tampering between members of different furrows over some time, until the respective furrow committees meet to try to resolve the matter. Even then, chances are that decisions may not be abided by or consensus may not be reached. There is currently no standing institutional mechanism to deal with interfurrow conflicts.

Other conflicts

Conflicts also exist between the irrigators and other water users. At the height of the dry season it is not uncommon for irrigation activities to leave no spring water for downstream use, be it for livestock watering, domestic uses, or other purposes. Relations with livestock keepers have generally been strained. Indeed, in critical situations livestock keepers have gone as far as destroying upstream irrigation-intake structures. Often, livestock graze close to the (unprotected) spring source. This practice compacts the soil, removes the vegetation, and denudes the surface. These processes threaten to reduce yield and quality of spring flow (Kimani 1988).

It is ironic that some of the indigenous Maasai are sitting on both sides of the fence: as landlords of irrigable land they benefit directly from irrigated cultivation and try to maintain or increase their access to water; at the same time, almost invariably they are livestock keepers and as such decry competition for water from irrigation activities.

Domestic water use also is affected by the conflict situation and comes under strain when irrigation does not leave any water for downstream use.

Gender issues

Women take an active part in all activities of the irrigated crop cycle, which includes land preparation, planting, weeding, water management, harvesting, produce grading and sorting, and local marketing. Indeed, women generally contribute more than men to the completion of these tasks (DIU 1993).

It is interesting to note, however, that when compared with rainfed farming, irrigated farming in Rombo on average gives women less of an unequal status and more of a say in management decisions. The reason for this is that irrigated farming is extremely labour intensive. Also, success and good profits depend on proper timing, careful coordination, and sound management of all the necessary inputs and tasks. Much more than in rainfed farming, men participate in working the fields, because they have little choice. Conversely, women tend to share in decision-making to a greater extent.

In Rombo, only a small number of active registered members of farmers associations in RIS (about 5%) (J.B.M. Kangethe, Rombo, personal communication, 1995) are women, most of them tenants. However, given the extensive experience in irrigated farming of many local women, they could and should play a much greater role in local water committees.

Those women who are active members of farmers' associations usually are de facto heads of household: either they are not married or their husbands work elsewhere in the country. These women are active irrigators and make most of the operational day-to-day decisions — waiting for a far-away husband to make decisions is not practical. This tends to put women in situations of greater responsibility and to lessen women's marginalization.

As for domestic water use, women and girls are the primary providers and are particularly affected by dwindling water availability and poor water quality.

Cultural and political change

Cultural aspects

RIS affords a good example of enormous cultural change that is taking place within the indigenous Maasai population. Historically, the Maasai always despised cultivation as a productive activity that seemed to them to be clearly inferior to nomadic pastoralism. However, in Rombo and in some of the other irrigation developments (and areas of rainfed cultivation) in Loitokitok Division, rapid transformation is leading to diversification among the Maasai away from a pure livestock economy toward agriculture and other investments like education.

The fact that it was people of Maasai origin who started irrigation activities in Rombo is likely to have removed some of the stigma attached to irrigated cultivation by the Maasai. Furthermore, intermarriage between the local Maasai and the many Kikuyu, Kamba, and other migrants who settled in Rombo after independence put the Maasai in greater touch with agriculture.

Nevertheless, Maasai generally preferred not to farm but to either sell or lease land to incoming migrants. However, in the last decade this has started to change, with more Maasai landowners taking up active farming themselves. The reasons for this change include the desire for greater profits and the realization that informal short-term leasing arrangements lead to land degradation and consume their land assets in the longer term. In Kisioki Furrow, of Upper Rombo, for example, 6 of the 16 Maasai landowner members are now active irrigators (H. Njoroge, former water distributer, Rombo, personal communication, 1995).

Political aspects

At the political level, recent multiparty politics (following the first multiparty elections in late 1992) and ethnic clashes elsewhere in Kenya have introduced a measure of ethnic rivalry and tension in Rombo (and elsewhere in Loitokitok). This tension has significantly affected irrigated cultivation. A group of Maasai, primarily young men, publicly decried the loss of land and other resources to other ethnic groups and advocated greater control by Maasai over "their" resources.

Notably, the group has attempted to "take over" irrigation committees in most irrigation schemes in Loitokitok Division. In a number of Rombo's canals, new Maasai-dominated committees were announced at meetings held in the second half of 1993. The newly appointed Maasai sought to assert themselves by promulgating new rules, such as "one farmer, one share." One of their main concerns has been ensuring access to land and irrigation water for sons of Maasai landlords.

Marketing problems

The Loitokitok area is the single largest source of horticultural produce in Kenya. The main markets are Mombasa and Nairobi. Total annual production of onions and tomatoes is in the neighbourhood of 3 000 – 4 000 t each. These figures are for 1988 (Kangethe 1990). More recent production levels may deviate from this. There are also pronounced fluctuations in annual production from year to year.

Farmers in Rombo and most other irrigation areas of Loitokitok Division face grave problems in marketing their produce. The two main roads (to Mombasa and Nairobi) are not paved or even all-season roads, and in the rainy season they become virtually impassable. There is not a single telephone connection in Rombo. Farmers have little, if any, market information that could inform their marketing decisions. It is not surprising that farm-gate prices are essentially dictated by middlemen and brokers, mostly based in Mombasa and Nairobi.

Competition from livestock and wildlife

There is considerable competition between livestock and irrigated agriculture for available water, as well as for land and grazing resources, in Rombo. Intensified and integrated modes of livestock production, such as irrigated fodder production and zero grazing, are still too sporadic and underdeveloped in the region to allow benefits to accrue from a more systematic integration of agriculture and livestock.

Wildlife has also caused stress on local water and land. In the long dry season (June–September), wildlife regularly enters the area from the bordering Tsavo National Park in search of water and food. In the process of reaching water sources, elephants, buffaloes, and other wild animals cause considerable damage to land, crops, and spring areas and inflict injury or even death upon humans.

Of course, wildlife brings enormous economic benefits through tourism in the park, but virtually all of these benefits accrue to others in Kenya and abroad. Much of the costs, however, are borne by local farmers. Wildlife damage detracts from surplus resources available to farmers that could be used for the conservation and improvement of water sources and structures and of land assets.

Water and land rights

Perhaps the most critical issue for current irrigation problems and successful future irrigation development in the Loitokitok region relates to property rights (water and land). Traditional resource property-rights systems are complex and work largely on an informal basis. Some of the indigenous Maasai people became formal individual owners of titled land at the time of land adjudication. They have sold part of this land in smaller parcels to a larger number of immigrants and have informally leased most of the remainder to immigrant tenants. Other Maasai were allocated land parcels on group ranch territory.

Acquisition of water rights has been contingent on owning or occupying land next to irrigation canals or on investing time and labour in irrigation systems. Those individuals who bought or inherited titled land or were allocated a parcel of collectively owned (such as group ranch) land within an irrigation area gained access to the in situ water resource as well. Similarly, those who helped construct or maintain an irrigation structure were entitled to a share of the water.

This resource tenure system may have worked reasonably well in the start-up and expansion phases of a particular irrigation system, but it has come under considerable strain as saturation of available water and land resources has occurred.

Informal land leasing

The large number of informal tenants presents a serious problem to sustainable and equitable resource use. Tenants have no incentive to conserve water and land resources or to maintain associated infrastructure. Their interest is usually in making the largest possible short-term profits with minimum inputs. Maasai landlords who lease out their land, on the other hand, are often not interested in farming and fear the possibility of tenants staking claims in land; they want to keep leasing arrangements informal and short term so that they will be able to evict tenants. This way, tenants and landlords collaborate in mining water and land resources for short-term benefits. Also, the tenants' profits are rarely, if ever, reinvested locally; rather, they tend to flow out of the area as remittances or in other forms.

To promote more responsible management of land and water resources, it is critical to promote long-term registered leasing arrangements. As well, there is a great need for awareness-building among Maasai landlords to avoid resource depletion on their lands. To some extent, this has already happened. This is also reflected in an improved ratio of landlords to tenants in Rombo. Ten years ago, there were nine tenants to one landlord (Maasai or non-Maasai); today, the ratio has dropped to 4 : 1 (Otieno and Njau 1995).

Informal irrigation-scheme development

Local irrigation activities and structures were initiated on the basis of personal initiative and later expanded as other farmers joined in, often paying the pioneer farmer for his initiative and investments. However, the informal nature of irrigation developments in Rombo and elsewhere in Loitokitok contributed to an uncontrolled mushrooming of irrigation activity. This has led to a situation where there is no clear sense of the rights of access to water for the different water users.

Until recently, water was essentially regarded as a "free" resource. Accordingly, water use and its distribution among different beneficiaries developed haphazardly. Only within individual irrigation-canal systems have reasonably effective self-regulating institutions evolved. Allocation of water across canals and between irrigation and other uses has been left to an ex-post balancing of interests and forces. No formal water permits have ever been issued for irrigation activities or other water uses. Thus, it is not surprising that water stress is considerable and conflict and tension run high.

Need for long-term approach

There is an urgent need for a careful, long-term approach to water allocation and a framework for water development that assigns different water entitlements to different competing users. Available water should be distributed on an equitable basis, after negotiations and with mutual agreement, in a participatory fashion. Such an approach would not be possible without a careful assessment of spring yields, total available water resources, and individual water needs. Also, irrigation intakes and canals would have to be upgraded, made permanent, and dimensioned as collectively agreed.

Summary and conclusions

The problem

Rombo's irrigation schemes have the virtue that they are developed and managed entirely by the farmers themselves. Unlike other "national" irrigation schemes, no subsidies are received from the government. Nevertheless, irrigation developments in Rombo (and in some other schemes elsewhere in Loitokitok Division) seem to be at a crossroads. Water stress and conflict have increased and reached critical levels in some of the canal systems and between irrigation and other water uses. Negative impacts are starting to be felt on crop yields and livestock productivity, the size of irrigated areas, and farming profits.

These problems have been a consequence of increasing water demand because of expanding numbers of farmers, inefficient irrigation technology, and lack of upgrading of irrigation structures; undefined or insecure property rights (water and land); and apparent declining overall water availability (lower and more variable spring yields).

This situation is exacerbated by severe marketing problems; damage to crops, land, and water sources by wildlife and livestock; opportunistic behaviour of most farmers and lack of local-level institutional cohesiveness; Maasai land owners' reluctance to farm; insufficient participation of women in local farmers' organizations; local political tension and infighting; outflow of farmers' cash surplus from the Rombo area for the benefit of outsiders; and widespread local resource degradation (both land and water).

The way forward

What would be needed to turn things around? First, water needs, rights, and responsibilities within irrigation and between irrigation and other uses need to be clarified. As well, land-tenure arrangements must be reexamined. Informal and short-term leasing arrangements need to be discouraged. At the same time, efforts need to be made to expand irrigation-water services through the gradual upgrading of irrigation structures, with outside assistance as desirable. However, such upgrading is not likely to happen unless the property-rights issue is dealt with effectively.

An overall framework and action plan for long-term equitable and sustainable water development should be worked out. This should include careful water resource assessment; awareness building, education, and training in sustainable forms of resource use, land leasing, and improved farming practices; development of effective participatory institutional mechanisms for transparent local water allocation and conflict resolution; and greater involvement of women in local decision-making processes and institutions.

This framework and action plan should build on existing structures and mechanisms and be developed, as much as possible, by the local people themselves and their representative institutions. It would be essential as well to mobilize the necessary local, regional, and national political support and to evolve effective modes of government assistance for local irrigation development and for the upgrading of general infrastructure (roads, telecommunications, and so on) in the area.

Finally, a number of research inputs would be required. At least the following areas would need priority attention: water-resource assessment; improvements in in-field water use and irrigation technology; mechanisms to upgrade irrigation structures; ways and means to raise awareness and provide skills, particularly among Maasai landowners, on improved irrigation and resources management; and the evolution (through participatory action research) of an equitable and sustainable resource property-rights system and self-regulating local institutions for resource management and conflict resolution.

References

DIU (District Irrigation Unit). 1993. District profile, Kajiado District, Rift Valley Province. District Irrigation Unit, Agricultural Engineering Division, Ministry of Agriculture, Livestock and Marketing, Kajiado District, Kenya. 56 pp.

Ewaso. 1994. Irrigation schemes in Loitokitok Division of Kajiado District. Report prepared for the Ewaso Ng'iro South Development Authority, Kenya. 43 pp.

Kangethe, J.B.M. 1990. Profitability of small holder farming systems with irrigation, Loitokitok Division. Report prepared for the ASAL Program Office, Kajiado District, Kenya. 108 pp.

Kimani, J.K., compiler. 1988. Evaluation of irrigation developments in Loitokitok Division. Report of a multidisciplinary task force. Rural Development Services. 29 pp.

PIU (Provincial Irrigation Unit). 1986. Rombo irrigation scheme, phase I: Maili Tatu and Ntere Kesi, Kajiado District. Report by the Provincial Irrigation Unit, Rift Valley Province, Land Development Division, Ministry of Agriculture, Kenya. 46 pp.

Rached, E. 1994. Report on the Pan-African Workshop on Water Management. Sponsored by the International Development Research Centre, 5–8 Dec. 1994, Cairo, Egypt. 30 pp.

NGO Experience, Intervention, and Challenges in Water Strain, Demand, and Supply Management in Africa

Oduor Ong'wen
Policy Analyst and Researcher, Bank Monitoring Unit, Africa Water Network, Nairobi, Kenya

Introduction

This paper examines the efforts of nongovernmental organizations (NGOs) and community groups to bridge the gap between human water needs and water supply in Africa. The paper looks at water stress, efforts to address water scarcity, and communities' expectations and coping strategies.

NGOs in the water sector are directly involved in mobilizing communities to improve their water-supply and management strategies by providing technical expertise and training. This paper reviews NGO intervention at the regional, national, and local levels, obstacles, and challenges.

Africa's stressed water systems

A "water stress index" based on the approximate minimum amount of water per person necessary to maintain an adequate quality of life in a moderately arid zone was developed by Falkenmark et al. (1989). The index established that about 100 L/d, or 36.5 m^3/year, is the minimum per person requirement for good health and that roughly 5–20 times this amount satisfies the requirements of agriculture, industry, and energy production. Hydrologists designate water-scarce countries as those with an annual freshwater availability of less than 1 000 m^3/person and water-stressed countries as those with an annual freshwater availability of 1 000 – 1 667 m^3/person. Although useful, such quantitative indicators can oversimplify the issue: they do not take into account important qualitative indicators, such as pollution levels in water sources or the distance people and animals have to walk to reach potable water. Of the 50 African countries, 9 are in the water-scarce category, 4 are in the water-stressed category, and at least 6 are at the brink of the water-stressed category. Per capita availability of renewable freshwater for the most water-stressed countries in Africa is shown in Table 1. Data on water in Africa, however, are both poor and incomplete.

Table 1. Annual per capita availability of renewable freshwater (1990).

	Available freshwater (m^3/person–year)		Available freshwater (m^3/person–year)
	Water-abundant countries		
Mauritius	2 047	Benin	5 625
Ethiopia	2 207	Niger	5 691
Lesotho	2 290	Cote d'Ivoire	6 177
Zimbabwe	2 312	Namibia	6 254
Nigeria	2 838	Mali	6 729
Tanzania	2 924	Chad	6 843
Burkina Faso	3 114	Zaire	27 253
Madagascar	3 331	Guinea Bissau	32 158
Togo	3 398	Sierra Leone	28 545
Ghana	3 529	Guinea	39 270
Uganda	3 759	Central African Republic	46 875
Mozambique	4 085	Equatorial Guinea	85 227
Mauritania	4 387	Liberia	90 097
Senegal	4 777	Gabon	141 501
Sudan	4 792		
	Water-stressed countries		
Libya	1 017	Egypt	1 123
Morocco	1 117	South Africa	1 317
	Water-scarce countries		
Djibouti	23	Algeria	689
Tunisia	540	Rwanda	897
Cape Verde	551	Somalia	980
Kenya	636	Malawi	939
Burundi	655		

The United Nations projects a further deterioration of the water-supply situation in the next three decades. The ever-widening gap between supply and demand has prompted some NGOs to get involved in the water sector.

The United Nations projects a further deterioration of the water-supply situation in the next three decades. The ever-widening gap between supply and demand has prompted some NGOs to get involved in the water sector.

Sources of stress

Increasing stress on Africa's freshwater resources is due to natural and human causes. Rapid population growth, pollution from pesticides and fertilizers, and industrial effluent all contribute to Africa's water stress. Another cause of stress is environmental degradation. Forests, which serve as important water catchment areas, are being cleared for fuelwood, lumber, and agriculture. Existing agricultural land is being degraded by soil erosion and devegetation.

NGO approaches

Most NGOs aim to empower local communities, rather than establishing large-scale water projects. These NGOs are devoted to strengthening the capabilities of community groups to use and manage water resources prudently and, through this action, to upholding local democracy and sustainable livelihoods. To be successful, projects involving civil society require an enabling environment to cement the delicate tissues of social relations. In the field of water, therefore, NGOs focus primarily on water and community participation or on water in support of civil society. Water and community participation stress integration through a number of approaches:

- A sustainable-development perspective entails the integration of the economy, ecology, health, and social well-being. NGOs in Africa see water, not as an end in itself, but as an entry point for sustainable community livelihood.
- The contribution of the intended beneficiaries through their voluntary associations, such as village water committees, is emphasized when projects and programs are implemented.
- Institutional arrangements are established to allow genuine participation of intended beneficiaries in decisions and actions aimed at improving their living conditions. Participation in local water management includes such aspects as arbitration, conflict resolution, design, implementation, monitoring and evaluation, technical assistance, resource mobilization, and contractual arrangements.
- Capacity-building and institutional development are considered essential for project or program sustainability.

- Consistency in intervention between local, national, regional, and (often) global levels is stressed.

Building on their comparative advantage over the state and private sectors, national, regional, and NGO networks play a fourfold role in the implementation of the approaches described above:

- supportive and catalytic agency for local initiatives;
- liaison and facilitator in forging alliances and networks to enhance exchange of information and experiences among different actors;
- mediator among the state, local communities, and external support agencies to encourage public intervention in the interest of disenfranchised populations; and
- educator and animator of civil society toward a sound, sustainable, and equitable use of water.

The water crisis and lack of financial resources available to community groups compel NGOs into cooperative arrangements centred on action and aimed at reduction of expenses. These collaborative arrangements include

- strengthening of national, regional, and international alliances and networks between NGOs;
- action-oriented projects focused on small-operation structures close to target populations; and
- democratic and transparent rules within contractual arrangements of obligations and accountability of participants.

Implementation

NGOs attempt to address water and civil-society issues through consistent intervention at all levels, which requires that they respond simultaneously to three interdependent objectives:

- water equity;
- decentralization; and
- local sustainability.

The three sections below elaborate these objectives and then suggest activities and strategies for meeting them.

Water equity

NGOs attempt to reform the approaches of international agencies to water issues by focusing on causes of impoverishment and fostering equitable patterns of water consumption and use.

1. *Activity*

   Awareness campaigns emphasizing the principles of justice, sustainable development, and peace on an ongoing basis

   Strategies

   Run national and regional campaigns funded by a partnership of stakeholders and aimed at raising awareness, understanding, and funds.

   Solicit the financial and technical support of bilateral and multilateral agencies involved in water, such as the United Nations Development Programme and the World Health Organization, to carry out campaigns.

2. *Activity*

   Public campaigns aimed at reforming water policies and programs of multilateral and bilateral agencies

   Strategies

   Promote cost-effective and self-sustaining financial mechanisms, including progressive water pricing systems, and better access to appropriate, low-cost credit and technologies for local communities.

   Develop cost-sharing models and projects among bilateral agencies, private-sector parties, and NGOs to involve communities in sustainable use of water at the community, regional, watershed, and riverbasin levels, as well as sustainable use of internationally shared waters.

   Redefine the international water component of the Global Environment Facility (GEF) within a more sustainable, integrated perspective of water, with funding accessible to local communities and with a participatory decision-making structure.

   Implement water and civil-society pilot projects, with NGOs as executing agencies and development assistance financing for a start-up fund and for international activities; in so doing, give priority to areas defined by the international agreements and programs.

3. *Activity*

Joint operations between NGOs and international actors on water and civil-society programs

Strategies

Set up consortia and other NGO groupings for the purpose of sharing technical, financial, and material means.

Implement pilot projects for sustainable exchanges of water, with NGOs as executing agencies.

Mobilize resources for water and civil-society operations through international associations of local governments and twin cities, the private sector, water professionals, and trade unions.

Promote participation of transnational companies in such operations.

4. *Activity*

Ongoing exchanges of information and experience sharing among NGOs of the North, South, and East on water and civil-society operations, on knowledge, techniques, and practices for sustainable water use, and on equitable institutional arrangements

Strategies

Systematize NGO participation in the negotiation and implementation of water-related aspects of various global environmental conventions (climate change and desertification) and programs (such as Capacity 21, GEF).

Fund regional meetings to assess and build on experiences of water and civil-society operations, to reorient collaborative arrangements, and to readjust existing multilateral programs toward such mutual learning.

Set up regional databases.

Decentralization

NGOs attempt to shift state intervention away from centralized, bureaucratic, sectoral approaches and toward the enabling water and civil-society approaches.

1. *Activity*

Delegation of water and civil-society operations to flexible, nongovernmental institutions close to communities

Strategies

Establish an institutional framework for decentralized control and management of water and sanitation services, which may be contracted to user associations.

Create regional project-funding banks to facilitate funding of water and civil-society operations.

Facilitate the access of local communities to the required inputs for such operations, including loans and credit; training; appropriate and low-cost technologies; and revolving funds generated from informal monetary flows, local savings, and community microbanks.

Provide technical and financial assistance to community groups to enable them to participate in municipal and regional water-related decision-making.

2. *Activity*

Application of national policies and programs related to sustainable and equitable use of water, integrating appropriate financing mechanisms

Strategies

Promote fair and efficient water- and sanitation-pricing systems with rates and fee-collecting mechanisms suited to disenfranchised populations; integrate the "polluter-pays" and "user-pays" principles.

Encourage the implementation of cost-sharing in water and sanitation operations, including, if required, support for contracts between the state, local communities, and user associations through an intermediary NGO, with financing modalities aimed at fostering sustainable capabilities.

Set up municipal and national trust funds based on a percentage of water fees and on savings resulting from the four R's (refuse, reduce, reuse, recycle) to finance community activities in poor neighbourhoods, with a focus on local user groups.

Support remuneration for local communities' intellectual property rights related to their expertise in local traditional water technologies.

3. *Activity*

Establishment of teams of community water-management practitioners to advise their local communities

Strategies

Finance training of these community practitioners within the contractual arrangement between state and educational institutions, acknowledging both community experience and new learning.

Recognize the community practitioner's status within the public utility, both as a resource for education and social mobilization and as a repository of meaningful expertise, and provide an appropriate system for remuneration (for example, through fellowships, grants, and community work).

Local sustainability

NGOs attempt to strengthen the local capabilities for initiating and maintaining sustainable patterns of water management and use.

1. *Activity*

Strengthening of local community organizations and reinforcement of their collective capabilities, with the aim of achieving sound and sustainable use of water

Strategies

Evolve banking-type mechanisms with flexible modalities assuring long-term loans with reasonable interest rates for water and sanitation operations, underwritten by NGOs or a third party.

Partition monetary flows in family and clan circles by setting up locally controlled microbanks as a way to tap local saving.

Build local revolving funds by various methods (such as those described above) to provide small-scale credits and loans; these funds could be facilitated by intermediary NGOs and should give priority to women.

Link local community groups to NGOs through long-term contractual arrangements and integrate institutional support costs, activity planning, mutual contributions, and ongoing learning.

Finance local practitioner and community knowledge and expertise in water management by regional community exchanges and a free system to permit replication and upgrading.

2. *Activity*

Consolidation of the legal status of local communities on land and water rights

Strategy

Support the consolidation of local communities' legal status to occupy and use land and water according to systems suited to local situations, such as free granting of land and water rights; spreading out payments and investing the funds in local development trust funds; and granting long-term leases.

3. *Activity*

Implementation of water and civil-society operations at the local level

Strategies

Provide technical and financial assistance to local project studies, calling upon the involvement of community specialists, to foster learning and self-sustainable capabilities.

Promote cost-recovery financing of water projects and infrastructures; cost recovery can be based on water-consumption fees, wastewater-treatment and reused-water savings, and reinvestment in community initiatives or on agreements with private-sector and city agencies to reinvest savings and part of the profits resulting from better water use in industrial parks.

Set up water services (distribution, sanitation, waste management) with diversified methods of community and other partner involvement, ranging from overall management to cost-sharing enterprises.

Involve exporting companies in funding small-scale productive activities recognizing the value of local expertise.

Compensate traditional and local expertise and technologies by locally appropriate methods through contributions to a common fund; provide advisory services.

NGO interventions

NGO approaches to water resources management are diverse. This section describes two programs to show the range of activities and orientations of NGO involvement in the water sector on the continent.

Epukiro Community Water Improvement Project (Namibia)

Government policy in Namibia is beginning to shift responsibility for water-point development to communities. The Epukiro Community Water Improvement Project (CWIP) was established to address the problems identified by the people and to develop their capacity to sustain and improve their water supplies (R. Yates, OXFAM, personal communication).

CWIP consists of six interrelated water projects to upgrade the availability, quality, and management of water in an area of small pastoral settlements. These six components are the following:

- technical advice;
- training for water committees and pumpers;
- rehabilitation of small dams;
- rehabilitation and hydrofracture of boreholes;
- installation of covered tanks and standpipes; and
- support for community initiatives in natural resources management.

In each community, project implementation is carried out through a village water committee and village elders. Requests and initiatives are made by these village committees to the water steering committee, which then selects beneficiaries. Thus, only active communities are included in the project, but encouragement and support are offered to help communities become active. Participatory rural appraisal techniques are used to help communities identify the need for and role of a water committee. Workshops and training courses are announced over the radio and through papers sent out with the monthly fuel ration, and the elders or the village water committee chooses whom to invite. The water steering committee also visits settlements to encourage participation.

Technical advice

Projects are coordinated by a water steering committee that represents the whole region. Project design and management, however, are handled jointly by this committee, a technical advisor, and a local counterpart. Additional management support is provided by the executive committee of the local farmer's association and by other community members through regular workshops.

Training for water committees and pumpers

Management and maintenance of diesel-powered pumps at boreholes are major problems in many communities. The engines are poorly maintained and out of action for days at a time because the operators employed by the community are not adequately trained and the government repair service lacks parts and transport. The community often does not understand that money spent on filters and maintenance will save money on repairs and fuel later. This can be overcome through training and education.

In places where there is no formal water committee, there would be an informal group that organizes fuel collection and purchase of spares. The group elicits contributions from people in the village, but on an ad hoc basis and without full consensus. Contributions are often set per household, not per animal, and the poorer farmers, recognizing the injustice, refuse to pay. The process creates tension in the village and causes delay in even the simplest repair. Again, training these communities on the benefit of a water committee and its possible role can overcome the problem. Committee treasurers are trained in appropriate book-keeping so that money or spare parts can be kept in reserve and simple repairs can be quickly effected.

Rehabilitation of small dams

The old excavation dams in the area are desilted by bulldozer, and the inlet canals are cleared by hand. This only takes place in settlements where the people have organized themselves and requested such help. Before work begins, a series of workshops are held to discuss ways of managing the dams in the future.

Rehabilitation and hydrofracture of boreholes

The boreholes themselves have had very little maintenance since they were drilled. Yields have fallen in many boreholes, even where the water table has not. In other cases, engines and pumps have been changed, with the result that boreholes are overpumped.

A program of borehole rehabilitation has begun, with a pilot scheme to find appropriate techniques. A sample of boreholes has been screened to select some for rehabilitation. After baseline pump-test and recovery-rate data are collected, a range of techniques is tried, starting with bailing and brushing, then jetting, hydrofracture, and chemical stimulation. After each technique, the hole is pump-tested again to see the effect. When complete, the Department of Water Affairs is informed of the sustainable yield so that the correct pump system can be installed.

The operators of these boreholes are given training and equipment to allow them to monitor yields and thereby take further action if they fail again. If possible, techniques are used that can be carried out in future by local drillers with percussion rigs.

Installation of covered tanks and standpipes

If the women of a village are active in the water committee and if the borehole water is of suitable quality, then a domestic water-supply point is installed. Water is pumped into a sealed tank that has been erected on a raised stand. The overflow flows into the livestock reservoir. One or two pipelines run from this tank to tap stands set at sites that have been agreed on by the water committee.

If the main boreholes are salty and suitable low-yielding, sweet-water sites are found, then handpumps are installed there instead.

To help mobilize the women's involvement in the project, a female extension officer has been employed and trained. She also encourages the women to look at other issues, such as sanitation.

Support for community initiatives

A fund is available to support other community initiatives throughout the region. Such initiatives can include small gardens, tree nurseries, education and training, workshops, and meetings. The fund is used to pay for preparatory work necessary to come up with future project proposals.

Mio–Gasera Water Supply Project (Ethiopia)

The Mio–Gasera Water Supply Project was initiated during the International Drinking Water Supply and Sanitation Decade, which emphasized community involvement in water supply and sanitation. The project has the highest level of community participation (A. Solomon, RADEN, personal communication).

The project covers a large geographical area, stretching 70 km from Mio village to Gasera town, in the province of Bale. The main source of income in the region is agriculture, predominantly livestock rearing. The aim of the project was to alleviate the acute shortage of potable water in the region. A spring was identified as the core of the project, and a 114-km pipeline, with a gravitational flow of 20 L/s, was constructed. This is the longest and biggest piped rural water-supply system in Ethiopia, serving about 65 000 people and their livestock.

The most outstanding feature of this self-help project was that local communities provided more than 65 000 person–days of unpaid labour. Before implementation, an ad hoc committee of representatives from the water agencies and the local administration drafted a document of participation guidelines. Regional party and government officials then mobilized the communities and provided digging tools. A cure for the chronic water shortages in the region was incentive enough for massive community participation in the implementation stage of the project. Most of the pipeline trench was dug in 3 months. Harder rock was excavated by skilled labour.

In Ethiopia's water sector, the water-development projects are usually monitored by a steering committee consisting of members from the finance, contractor, and owner organizations, and, sometimes, water-committee members. Such organizations usually work in isolation. In this project, however, participation came from all quarters: the local administration handled the construction; the head of community participation mobilized and educated the communities; and UNICEF, the donor, supplied all the necessary equipment. Education materials on water use, basic sanitation, and personal hygiene were prepared and distributed to relevant extension agencies for project use. The local development agencies constantly followed up on the safe utilization and maintenance of the water schemes. It was a collaborative effort. External funds covered 70–80% of the water-supply-sector investment. The remaining funds came from government and the community.

Many rural water-supply schemes fail to be sustainable because of lack of strong community participation, inadequate funds for operation and maintenance, inappropriate technology, and untrained technicians. However, this project has proven to be sustainable and is functioning as expected and designed. The existing policy is that the community should at least cover the operation and maintenance cost. The local community contributes some money regularly through the existing water committee to cover the cost of minimal operation and minor maintenance, and the water-supply unit of the region handles the cost of major maintenance.

The scheme is not fully managed by the community, but the water committee still plays an important role. Its main responsibilities are to collect water bills from the community and safeguard the water scheme. The water-supply unit of the region takes care of the whole water-supply scheme and is responsible for overall operation and maintenance, administration, and coordination of beneficiaries through the water committee.

The local economy has grown considerably since the completion of the project, and the beneficiaries realize the importance of self-help. They now have adequate and reliable water, disease outbreaks have decreased because of the cleaner water and better hygiene, and women spend less time collecting water.

The absence of effective coordination of development programs and of logistic support for community participation, plus shortages in construction materials, hampered the smooth operation of construction activities. However, the community has developed a sense of attachment to the water scheme because it is the product of their labour. The people feel responsible for maintaining the standards of operation. As a result, the scheme survived the political instability of 1991 when other projects fell apart. In addition, the participation guidelines and the subsequent formation of coordinating committees have greatly assisted in carrying out community project activities effectively and in a coordinated manner.

A number of lessons can be learned from this experience:

- A large water-supply project like the Mio–Gasera one, which is gravity fed and involves large-scale participation of communities, is technically and economically feasible and worthy of being replicated elsewhere.
- It is important to integrate a water-supply development program with other local development-program components, such as agricultural services, home economics, health education, low-cost sanitation, and adult education programs.
- Local beneficiaries should be given the chance to identify their problems and, as much as possible, participate in planning, implementing, operating, maintaining, monitoring, and evaluating any rural project.
- Special consideration and effort should be made to mobilize women's participation and thus to enhance sustainability.

NGOs and research

Research is not a priority for NGOs. As can be seen from the case studies, many NGOs began their activities in the water sector through direct action in water-supply projects. Most African NGOs are externally funded, and most donors are more concerned about water-supply issues. Therefore, few resources are available for research.

Within the limits of their resources, NGOs are participating in research in such areas as appropriate technology, traditional coping mechanisms, rainwater harvesting, pollution control, and source protection. Such research is carried out by individual NGOs or through collective efforts. An example of this collective approach to simple research is the Participatory Learning Network, created by Kenyan NGOs and based at the Network on Water and Sanitation in Nairobi. Important though it may be, if NGO participation in water research is to increase, more resources will have to be mobilized.

Future projections

NGOs and community groups are focusing their attention on alternative water-management strategies in the following areas: information; institutional development and capacity-building; promotion of popular participation; and sustainability.

Information

NGOs have recognized the vital role that dissemination and information exchange play in promoting sustainable use and management of water resources. They are, therefore, exploring ways and means of creating national networks and strengthening existing ones. This will facilitate access by NGOs and other social actors in the water sector to other experiences, successes and failures, and new ideas for dissemination.

Institutional development and capacity-building

Institutional capacity-building and the empowerment of community groups, particularly marginalized groups, are primary objectives for many NGOs in Africa. They therefore aim at mobilizing human and material resources for training in management of information and resources, lobbying skills and policy analysis, and technical skills.

Promotion of popular participation

Regional NGOs recognize that many policymakers are increasingly integrating the concept of community participation into their policies, programs, and projects. However, the level of participation is often symbolic, and beneficiaries are not fully empowered. A number of regional NGOs are working to develop standards and indicators for evaluating successes or failures of projects; full community participation will be one of the key indicators. NGOs are also focusing on gender roles as a crucial element in water supply and management. In addition, they are already giving support to the continuing formation of water committees and to research on traditional values and knowledge, with a view to adapting them to suit current socioecological conditions.

Sustainability

The NGO agenda for ensuring future sustainability, as developed in an action plan at the African Water Network Workshop in June 1994, includes the following:

- developing holistic educational programs to increase the understanding, at the grass-roots level, of sustainable development;
- promoting efficient and effective use of water, as well as simple, affordable, and appropriate technologies;
- promoting techniques in critical self-evaluation and training communities in their use;

- encouraging communities to address the economic aspects of water supply and management by providing improved options, such as the establishment of water kiosks, locally appropriate pricing, and maintenance and water costing;
- ensuring public accountability and transparency at all levels and developing good resource-management procedures; and
- encouraging support agencies to downplay their traditional short-term project approach and promote long-term sustainable programs.

The agenda will be used by NGOs at all levels in Africa to obtain the objectives of equity, decentralization, and sustainability in water supply, management, and use. However, the approaches will remain focused on community participation. In such areas, NGOs will continue to have an important, indeed an essential, role to play in improving the quality of life and reducing stress on water resources in Africa.

Reference

Falkenmark, M.; Lundqvist, J.; Widstrand, C. 1989. Macro-scale water scarcity requires micro-scale approaches. Natural Resources Forum, 13(4).

APPENDIX I: WORKSHOP PARTICIPANTS

Ahmed E. Belal
Allaqi Research Group
Faculty of Science
Aswan Branch
Assuit University
Aswan, Egypt
Tel: [20] (97) 480 446
Fax: [20] (97) 480 450

Jean Biémi
Lecturer
Faculté des sciences et technique
Université d'Abidjan
22 BP 582
Abidjan 22, Côte d'Ivoire
Tel: [225] 44-90-00
Fax: [225] 44-04-12

Gérald Bourrier
Director
Regional Office for West and Central Africa
IDRC
PO Box 11007, CD Annexe
Dakar, Senegal
Tel: [221] 24-42-31
Fax: [221] 25-32-55

David Brooks
Program Director
Environmental Policy Program
Environment and Natural Resources Division
IDRC
PO Box 8500
Ottawa, ON, Canada K1G 3H9
Tel: (613) 236-6163
Fax: (613) 238-7230

Gilles Cliche
Senior Program Officer
IDRC
4 El Hesn Street
PO Box 14
Orman, Giza, Cairo, Egypt
Tel: [20] (2) 573-8926
Fax: [20] (2) 571-7690

Diana de Treville
Regional Representative
Winrock International
Institute for Agricultural Development
Box 45233
Nairobi, Kenya
HQ: Route 3, Morrilton, AR 72110, USA
Tel: (501) 727-5435
Fax: (501) 727-5411

Bassirou Diagana
09 BP 398
Ouagadougou 09, Burkina Faso
Tel: [226] 25-25-85
Fax: [226] 31-03-61 c/o CREPA Ouagadoudou

Samiha Fahmy El-Katsha
Senior Consultant
Social Research Centre
American University in Cairo
113 Kasr El Aini Street
Cairo, Egypt
Tel: [20] (2) 354-2964
Fax: [20] (2) 355-7565

Hosny El-Lakany
Director
Desert Development Centre
American University in Cairo
113 Kasr El Aini Street
Cairo, Egypt
Tel: [20] (2) 357-6331
Fax: [20] (2) 355-7565
Direct fax/tel: [20] (2) 357-6330

Osman Fadl
Fresh Water Consultant
CEDARE
21/23 Giza Street
Nile Tower Building
PO Box 52
Orman, Giza, Cairo, Egypt
Fax: [20] (2) 570-3242
Tel: [20] (2) 570-3473

Simon Forster
Economic Project Evaluation (Pty) Ltd
PO Box 563
Rivonia 2128, South Africa
Tel: [27] (11) 462-4405
Fax: [27] (11) 704-3944

Salwa Gomaa
Associate Research Professor
Social Research Centre
American University in Cairo
113 Kasr El Aini Street
Cairo, Egypt
Tel: [20] (2) 357-6792
Fax: [20] (2) 355-7565

Abdou Hassan
Ministry of Hydraulics and Environment
BP 257
Niamey, Niger
Tel: [227] 73-20-25
Fax: [227] 73-35-88

Davies Humphrey
Representative
Ford Foundation
Cairo, Egypt
Tel: [20] (2) 355-2121
Fax: [20] (2) 355-4018

Kwamé Karikari
Senior Lecturer
School of Communication Studies
PO Box 53
University of Ghana
Legon, Ghana
Tel: [233] (21) 774693
Fax: [233] (21) 224935

George Khroda
Director
Research and Programs
Red Plan Consultants Ltd
PO Box 56745
Nairobi, Kenya
Tel: [254] (2) 713593
Fax: [254] (2) 339155

Janusz Kindler
Senior Water Resources Planning Specialist
Europe and Central Asia Region; Middle East and North Africa Region
World Bank
1818 H Street NW
Washington, DC 20433, USA
Tel: (202) 473-2253
Fax: (202) 477-1981
e-mail: jkindler@worldbank.org

Fawzy Kishk
Director
Regional Office for Middle East and North Africa
IDRC
4 El Hesn Street
PO Box 14
Orman, Giza, Cairo, Egypt
Tel: [20] (2) 573-8926
Fax: [20] (2) 571-7690

Mohammed S. Matoussi
Resource Economist
Faculté des sciences économiques et de gestion
Université de Tunis
1060, avenue 7 novembre
Tunis, Tunisie
Tel: [216] (1) 514-303
Fax: [216] (1) 502-222

Adama Mbaye
Ministère de l'Hydraulique
Direction de l'hydraulique et de l'assainissement
BP 2041
Dakar, Sénégal
Tel: [221] 32-42-79

Nigel E. Motts
Project Officer
Regional Office for Southern Africa
IDRC
PO Box 477
WITS 2050, South Africa
Tel: [27] (11) 403-3952
Fax: [27] (11) 403-1417

Charles M. Namafe
Lecturer II
University of Zambia
Lusaka, Zambia

Maurice M. Ndege
Lecturer
Department of Civil Engineering
University of Nairobi
PO Box 30197
Nairobi, Kenya

P. Ngpipandulwa
Director
Rural Water Supply
Department of Water Affairs
Private Bag 13193
Windhoek, Namibia
Tel: [264] (61) 296-3061
Fax: [264] (61) 232-581

Arthur William Obare
Ministry of Land Reclamation
Regional and Water Development
Nairobi, Kenya

Osita Ogbu
Senior Regional Program Officer
Economic Policy
Regional Office for Eastern and Southern Africa
IDRC
PO Box 62084
Nairobi, Kenya
Tel: [254] (2) 713160
Fax: [254] (2) 711063
e-mail: oogbu@idrc.ca

Oduor Ong'wen
Policy Analyst and Researcher
Bank Monitoring Unit
African Water Network
PO Box 10538
Nairobi, Kenya
Tel: [254] (2) 556943
Fax: [254] (2) 555513
E-mail: awn@elci.gn.apc.org

Luke Onyekakeyah
Environmental Scientist and Consultant
PO Box 26522
Nairobi, Kenya
Fax: [254] (2) 230145

Eglal Rached
Senior Program Specialist
IDRC
4 El Hesn Street
PO Box 14
Orman, Giza, Cairo, Egypt
Tel: [20] (2) 573-8926
Fax: [20] (2) 571-7690

Eva Rathgeber
Director
Regional Office for Eastern and Southern Africa
IDRC
PO Box 62084
Nairobi, Kenya
Tel: [254] (2) 713160
Fax: [254] (2) 711063
e-mail: erathgeb@idrc.ca

Maurice Saade
Program Officer
IDRC
4 El Hesn Street
PO Box 14
Orman, Giza, Cairo, Egypt
Tel: [20] (2) 573-8926
Fax: [20] (2) 571-7690

Abdin M.A. Salih
Programme Specialist
Hydrology Sciences and Ecology
ROSTAS/Cairo
Unesco
8 Abdel Rahman Fahmy Street
Garden City, Cairo 11511, Egypt
Tel: [20] (2) 354-3036
Fax: [20] (2) 354-5296

Allain Sawadogo
Faculté des sciences et technique
Laboratoire d'hydrogéologie
BP 7021
Ouagadougou 03, Burkina Faso
Tel: [226] 36-30-38

M.P. Sesoko
DBSA
South Africa
Tel: [27] (11) 313-3086
Fax: (27) (11) 313-3264

John Sinclair
Development Counsellor
Canadian Embassy
Cairo, Egypt
Tel: [20] (2) 354-3119
Fax: [20] (2) 356-3548

Olanrewaju Smith
Program Officer
Environment and Natural Resources
Regional Office for West and Central Africa
IDRC
PO Box 11007, CD Annexe
Dakar, Senegal
Tel: [221] 24-42-31
Fax: [221] 25-32-55

Firouz Sobhani
Chief
Technical Unit
UNSO–UNDP
1 United Nations Plaza, Room FF-906
New York, NY 10017, USA
Tel: (212) 906-6614
Fax: (212) 906-6345

M.S. Solh
Regional Coordinator
ICARDA–NVRP
PO Box 2416
Cairo, Egypt
Tel: [20] (2) 572-8099
Fax: [20] (2) 572-4358

Geoffrey Stiles
Energy Conservation Advisor
SADC Energy Management Project
PO Box 949
Harare, Zimbabwe
Tel: [263] (4) 735-795
Fax: [263] (4) 791-202

Marc Van Ameringen
Director
Regional Office for Southern Africa
IDRC
PO Box 477
WITS 2050, South Africa
Tel: [27] (11) 403-3952
Fax: [27] (11) 403-1417

Joachim Voss
Program Director
Sustainable Production Systems
Environmental Policy Program
Environment and Natural Resources Division
IDRC
PO Box 8500
Ottawa, ON, Canada K1G 3H9
Tel: (613) 236-6163
Fax: (613) 238-7230

About the Institution

The International Development Research Centre (IDRC) is committed to building a sustainable and equitable world. IDRC funds developing-world researchers, thus enabling the people of the South to find their own solutions to their own problems. IDRC also maintains information networks and forges linkages that allow Canadians and their developing-world partners to benefit equally from a global sharing of knowledge. Through its actions, IDRC is helping others to help themselves.

About the Publisher

IDRC Books publishes research results and scholarly studies on global and regional issues related to sustainable and equitable development. As a specialist in development literature, IDRC Books contributes to the body of knowledge on these issues to further the cause of global understanding and equity. IDRC publications are sold through its head office in Ottawa, Canada, as well as by IDRC's agents and distributors around the world.